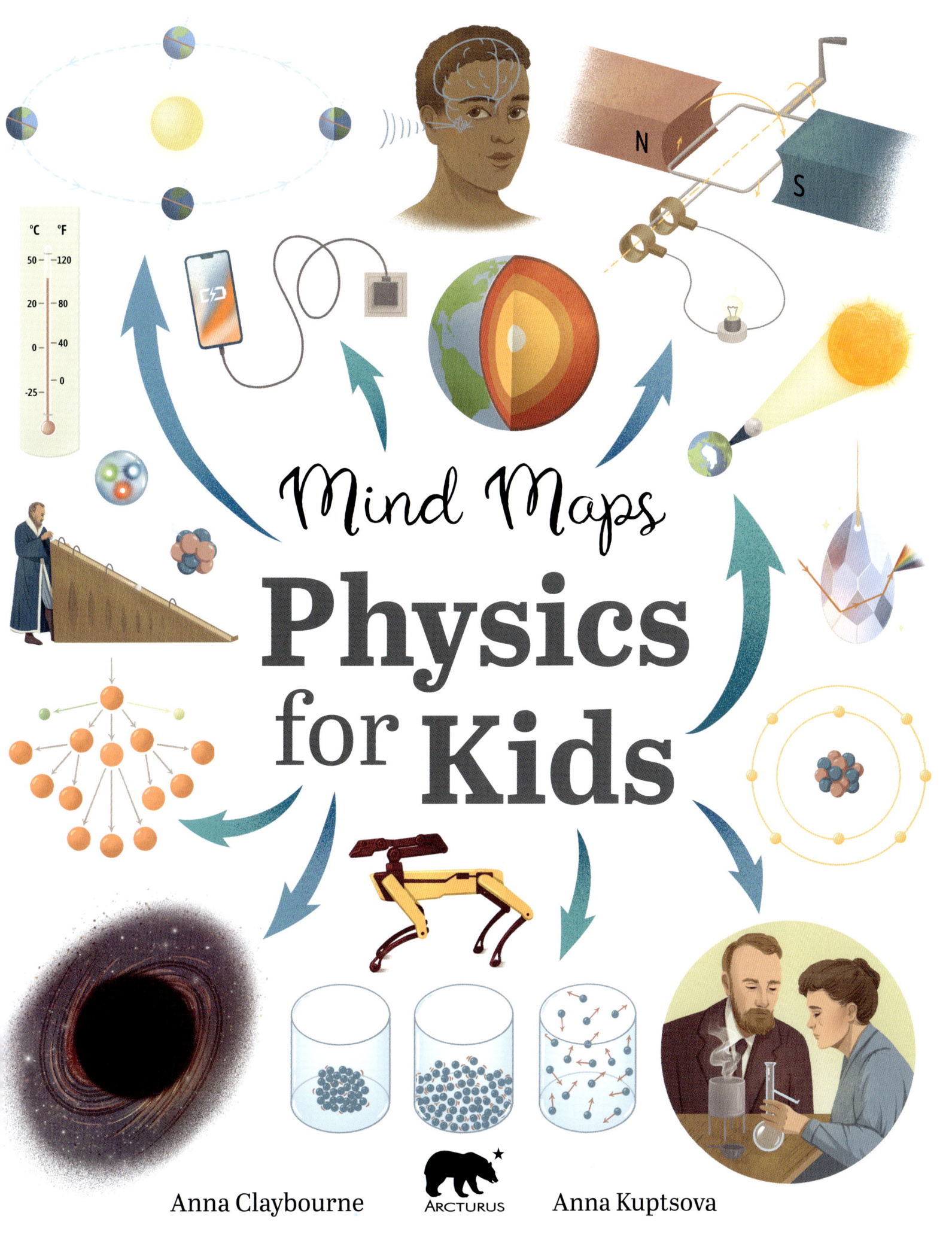

This edition published in 2024 by
Arcturus Publishing Limited
26/27 Bickels Yard, 151–153 Bermondsey Street, London SW1 3HA

Copyright © Arcturus Holdings Ltd

Anna Claybourne has asserted her right to be identified as the author of this text in accordance with the Copyright, Designs, and Patents Act 1988.

All rights reserved. No part of this publication may be reproduced, stored in a retrieval system, or transmitted, in any form or by any means, electronic, mechanical, photocopying, recording, or otherwise without prior written permission in accordance with the provisions of the Copyright Act 1956 (as amended). Any person or persons who do any unauthorized act in relation to this publication may be liable to criminal prosecution and civil claims for damages

Author: Anna Claybourne
Consultant: Anne Rooney
Illustrator: Anna Kuptsova
Designer: Sally Bond
Design Manager: Rosie Bellwood-Moyler
Editor: Lydia Halliday
Editorial Manager: Joe Harris

ISBN: 978-1-3988-2854-4
CH011051US

Supplier 29, Date 1024, PI 00007717

Printed in China

How this book works

In science, everything is connected!

Physics is the science of matter and energy. Matter means stuff—the stuff that makes up everything. Energy is what makes matter move, change, or do things.

For example, an apple is made of matter.

When it falls from a tree, it has movement energy.

Physics covers a wide range of topics, such as ...

- Electricity
- Heat
- Motion, or movement
- Light
- Atoms
- Sound
- Forces such as gravity
- Waves
- Stars, planets, and orbits

Just like other areas of science, physics topics are all connected and linked.

For example, think of **sound**, one of the forms (or types) of energy. This topic links to many others in physics, such as ...

- The way sound travels as a type of wave
- How sounds make molecules in the air move
- How sound can make matter heat up
- The way different types of motion, or movement, create different sounds
- How sound can't travel in a vacuum (empty space)
- How a flow of electricity makes a speaker or electric guitar play sounds

Make the link ...

This book links topics together in a big mind map, so you can see how everything is connected. Whenever there's a connection to a related topic, you'll find a link, like this:

You can read the different sections and topics in any order, or just choose a topic you're interested in, using the contents on the next page. Then follow the links to find out more, or jump to a different page and explore!

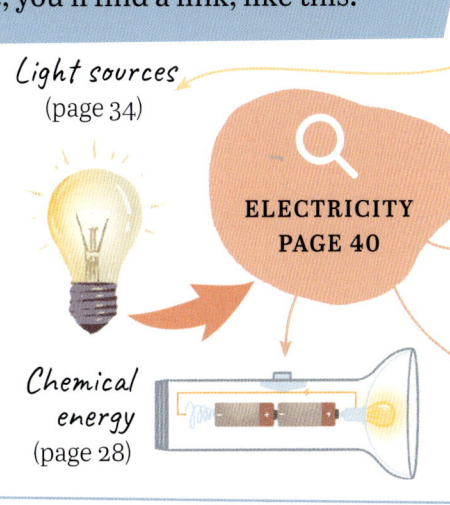

Light sources (page 34)

ELECTRICITY PAGE 40

Chemical energy (page 28)

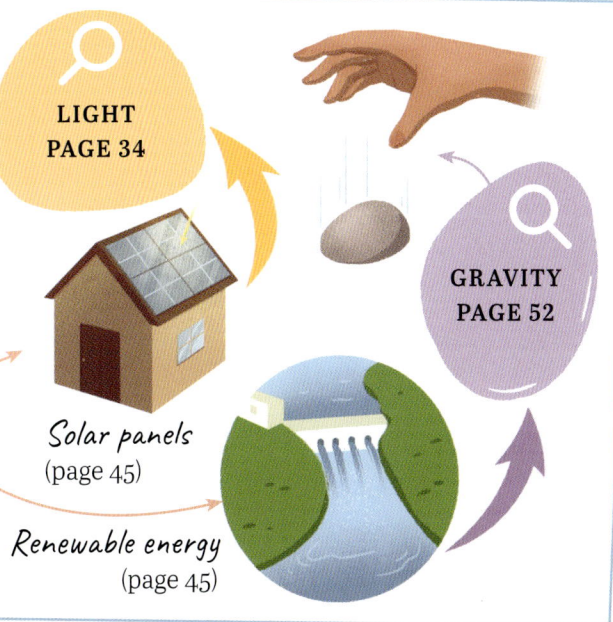

LIGHT PAGE 34

GRAVITY PAGE 52

Solar panels (page 45)

Renewable energy (page 45)

What's in the book?

Introduction: The world of physics

How matter works	6
Physics and physicists	8
Making measurements	10
How it started	12
Matter matters	14

Energy

What is energy?	16
Kinetic energy	18
Potential energy	20
Heat energy	22
Heat transfer	24
Heat expansion	26
Chemical energy	28
Energy waves	30
Sound	32
Light	34
Reflection and refraction	36
The electromagnetic spectrum	38

Here are all the Physics topics you'll find in this handy book.

Use the page numbers to find the topic you want, or just dip in at random!

Electricity

What is electricity?	40
Electric circuits	42
Generating electricity	44
Electricity supplies	46
Electronics	48

Forces and motion

Pushes and pulls	50
Gravity	52
Friction	54
Springs and elastic	56
Pressure	58
Magnetic and electric forces	60
Laws of motion	62
Measuring motion	64
Simple machines	66

Look out for the links to different topics on each page, like this:

ELECTRICITY PAGE 40

Matter

What is matter?	68
States of matter	70
Radiation and radioactivity	72
Fission and fusion	74
Mysteries of matter	76

Earth and space

Geophysics	78
The Universe and space	80
Shining stars	82
In orbit	84
Studying space	86
Mysteries of the Universe	88

Physics super mind map	90
Glossary	92
Index	94

How matter works

Matter is the stuff that everything is made of. Planets and moons, houses, cars, furniture, books, water, air, plants, animals, and our own bodies are all made of matter.

Physics is about how matter behaves—for example, how it moves, changes shape, speeds up, slows down, or gets hotter or colder. These things can only happen if energy makes them happen, so physics is about energy, too.

This bike, like all other objects, is made of matter.

Plastic seat
Metal frame
Rubber tires

Chain moves back wheel
Pedals move chain
Bicyclist pushes on pedals
Wheels spin
Bike moves

There are many different types of matter, energy, and movement. Physics explores all of them and how they interact with each other.

The bicyclist uses energy from her body to push on the pedals, making the bike move forward.

Types of matter

Matter comes in a huge variety of types that behave in different ways.

For example, metals such as iron, copper, silver, and tin have these properties, or features:

They can usually be bent and reshaped, like this steel paper clip.

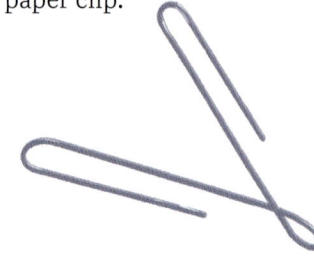

They conduct (or carry) electricity easily. That's why electric wires are made of metal.

They conduct heat well, too.

Most metals melt only at very high temperatures.

Pans are usually made of metal because the heat spreads quickly through the base. Their high melting point means they don't melt on the stove.

Each type of matter has its own properties and abilities.
- Rock — High melting point
- Water — Heats up slowly
- Wood — Strong, but floats
- Diamond — Extremely hard and transparent
- Plastics — Light and waterproof
- Rubber — Tough and flexible

🔍 HEAT TRANSFER PAGE 24

🔍 MATERIALS PAGE 68

Different states

Matter can exist in three main states: solid, liquid, and gas. Usually, the state matter is in depends on its temperature.

When water is very cold, below 0°C (32°F), it freezes into solid ice.

Between 0°C (32°F) and 100°C (212°F), it's a liquid.

Above 100°C (212°F), water boils and becomes a gas.

STATES OF MATTER PAGE 70

Matter and forces

All the time, forces push or pull on matter, making it do different things.

Friction is an example of a force. It makes things slow down, grip, or stop when they scrape or rub together.

A bike's brakes rub against the wheels, making then slow down, then stop.

FRICTION PAGE 54

Understanding energy

Every time something happens to matter, energy converts, or changes, from one form to another.

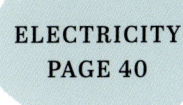

ELECTRICITY PAGE 40

For example, when you switch on a lamp, electrical energy changes into light energy.

Physics and physicists

Scientists who study physics are called physicists. There are thousands of them around the world, working on many different areas of physics.

This physicist is doing an experiment using sound waves to make small beads levitate, or float in midair.

Levitating bead

Besides doing experiments in labs, physicists can work in a wide range of different places and jobs.

Physicists often use hi-tech equipment to do experiments. This experiment uses a special speaker called an acoustic levitator. It releases high-energy ultrasound, which is too high-pitched for humans to hear, but can be directed at objects to hold them up.

Speaker creates sound waves

Levitating bead

This part reflects sound waves

Areas of physics

Physics has many different branches and areas, such as:

- Newtonian mechanics — Forces and motion in large and everyday-sized objects.
- Thermodynamics — Heat energy, temperature, and movement.
- Geophysics — Movements and energy of parts of the Earth.
- Particle physics — The smallest particles that matter is made of.
- Astrophysics and cosmology — Stars and space objects, and how the Universe works.
- Theoretical physics — Ideas and theories about physics.
- Physical oceanography — Movements and energy of the seas and oceans.
- Quantum mechanics — How very small particles move and behave.

PARTICLES PAGE 68

THE UNIVERSE PAGE 80

College and university physics

Colleges have big physics labs filled with tools and equipment. Physicists who work here have two main jobs:

They teach physics to students and show them how to do lab experiments ....

This university physicist is studying waves in liquids.

... and they also do their own research and experiments, too, to come up with new discoveries or inventions.

WAVES IN LIQUIDS PAGE 30

SPACE TRAVEL PAGE 87

Applied physics

Studying physics is interesting in itself, but it also has many important uses. Physicists often work for organizations and companies, using their skills to design or make things.

For example, an acoustician, or sound physicist, could design concert halls or sound effects.

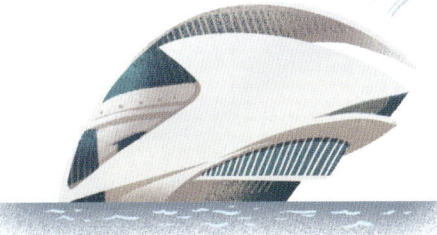

Spain's Palau de les Arts Reina Sofía opera house

And some physicists work for space agencies, planning and calculating the routes for rocket launches and space missions.

Theoretical physics

This type of physics doesn't need much equipment at all—just a pencil and paper or a computer. Theoretical physicists use mathematical calculations and equations to figure out how things work and make predictions.

For example, space scientists first detected black holes in 1971. But long before that, theoretical physicists had predicted that they probably existed, because of the way matter and gravity work.

John Michell was one of the first scientists to predict black holes, in the 1780s.

BLACK HOLES PAGE 88

9

Making measurements

Physicists observe how things move and change, and they also do lots of calculations. So they often need to measure and count things using units of measurement.

For example, physicists sometimes do tests on graphene, a very light, strong material, to find out how it can be used.

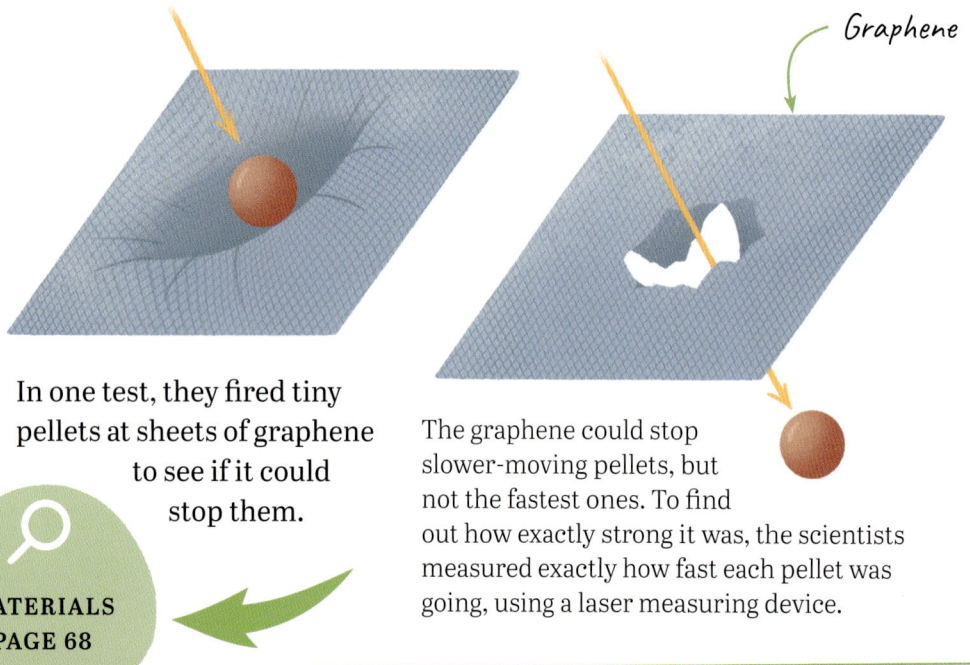

In one test, they fired tiny pellets at sheets of graphene to see if it could stop them.

The graphene could stop slower-moving pellets, but not the fastest ones. To find out how exactly strong it was, the scientists measured exactly how fast each pellet was going, using a laser measuring device.

To make measurements like these, physicists need units of measurement. They use different units for different things.

MATERIALS PAGE 68

MASS AND WEIGHT PAGE 53

Space and distance

The most basic units of measurement measure things like length or distance, area, and volume. In physics, units are often related to each other.

- The basic unit of metric **length** is the meter (m).
- There are 100 centimeters (cm) in a meter,
- and 10 millimeters (mm) in a centimeter.
- A kilometer is 1,000 meters.
- **Area** is measured in square mm, cm, m, or km.

- **Volume** is measured in cubic mm, cm, m, or km.
- A liter (l) of water is a cube of water 10cm x 10cm x 10cm.

One cubic cm is a milliliter (ml), or 1/1,000 of a liter.

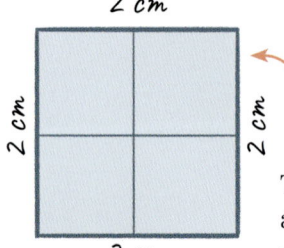

This box has an area of 4 cm², or square centimeters.

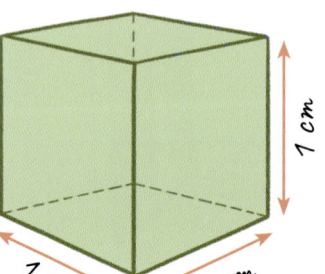

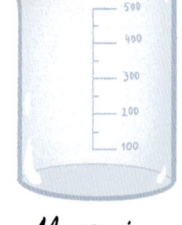
Measuring beaker

Digital lab clock

- **Mass** (the amount of matter in something) is measured in grams.
- One milliliter of water has one gram of mass.
- A kilogram (kg) is 1,000 grams.

Measuring time

Time is measured in units based on the movements of the Earth.

- A year is the time it takes the Earth to make one orbit of the Sun.
- A day is the time it takes the Earth to spin all the way around.
- There are 365.2422 days in a year,
- 24 hours in a day,
- 60 minutes in an hour,
- and 60 seconds in a minute.

By combining units of distance and time, you can measure speed. For example, if a train is moving at 200 km/h (124 mph), it travels 200 km (124 mi) in one hour.

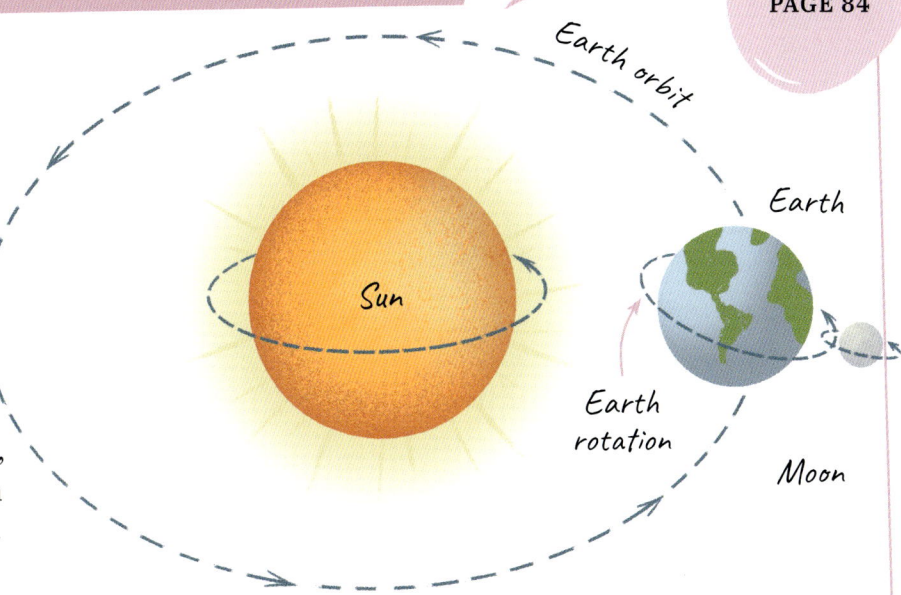

ORBITS PAGE 84

More measurements

There are many more units of measurement, too. Here are some of them...

Unit	Used to measure...
Joule	Energy
Newton	Force
Pascal	Pressure
Ampere	Electric current
Ohm	Resistance in an electric circuit
Watt	Power
Candela	Brightness of light
Hertz	Frequency
Degrees Celsius	Temperature
Degrees Fahrenheit	Temperature
Kelvins	Temperature
Decibel	Intensity of sound

LIGHT PAGE 34

Measuring equipment

Since measuring is so important, physicists need very accurate, reliable measuring equipment.

For example, to measure mass, they use a machine called a lab balance or analytical balance. It's similar to a kitchen weighing scale, but much more sensitive.

TEMPERATURE PAGE 23

How it started

For thousands of years, people have been wondering about what everything is made of, studying the stars, and using machines to control forces.

Ancient civilizations around the world made several physics discoveries and inventions.

- 3500 BCE — Sumerians invented the first known wheeled vehicles.
- 2600 BCE — Indus Valley people invented a measuring system.
- 2500 BCE — Ancient Egyptians used ramps to make building pyramids easier.
- 500 BCE — Olmec people learned to predict the motion of the Sun, Moon, and planets.

Through the centuries, scientists made more and more physics discoveries, and built up a vast amount of physics knowledge.

This is a ruler found in the Indus Valley city of Mohenjo-Daro.

🔍 RAMPS PAGE 66

Ancient Greeks

The ancient Greeks are famous for their scientific experiments and ideas.

Several Greek thinkers had theories about what matter was made of. Around 400 BCE, Democritus said that it was made of tiny units in different types and sizes, which he called atoms—an idea that later turned out to be true.

Democritus

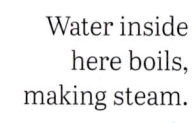

Atoms

Steam shoots out here, making the ball spin.

Water inside here boils, making steam.

The inventor Hero of Alexandria built the first steam engine, which changed heat energy into a spinning movement.

Islamic golden age

From the 700s to the 1200s, the city of Baghdad (in what is now Iraq) was a center of learning, where Islamic scientists studied and shared ideas.

Astronomers collected and wrote down knowledge of the stars, and gave them names in Arabic. Many stars still have their Arabic names today.

Fomalhaut

Fomalhaut is a star in the Southern Fish constellation. Its name comes from the Arabic phrase "fam al-ḥūt, meaning "the whale's mouth."

STARS PAGE 82

Giants of physics

Galileo Galilei and Isaac Newton, who lived in the 1500s and 1600s, were among the greatest physicists of all time.

Galileo experimented with forces and motion, and built a telescope to study the Moon and planets.

In one experiment, Galileo is said to have dropped metal balls off the Leaning Tower of Pisa to show that lighter objects do not take longer to fall.

Newton explained how gravity and orbits worked, as well as studying motion, light, color, and sound.

Newton studied the way a prism can refract or bend light, splitting it into the colors of the rainbow.

REFRACTION PAGE 37

Understanding atoms

In the twentieth century, physicists such as Neils Bohr and Ernest Rutherford did experiments that revealed the structure of atoms and the parts they were made of.

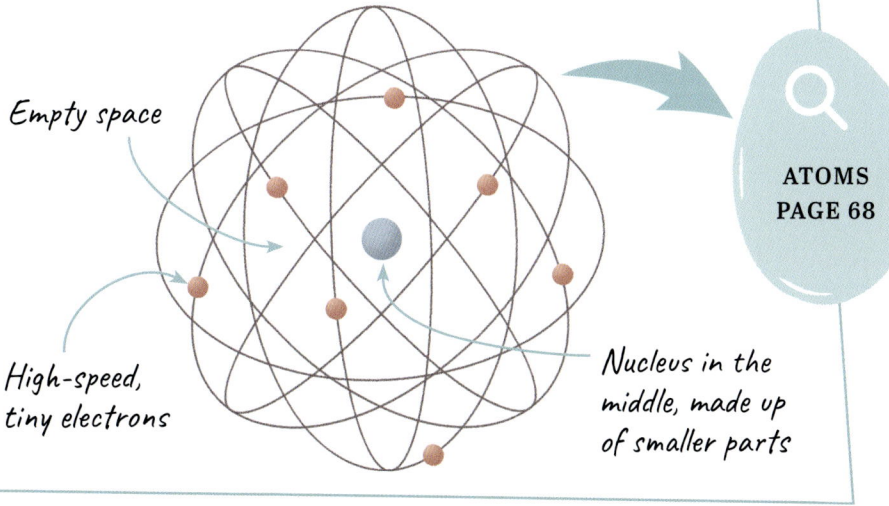

Empty space

High-speed, tiny electrons

Nucleus in the middle, made up of smaller parts

ATOMS PAGE 68

13

Matter matters

Physics is not just a science, but also a central part of everyday life. All our machines and gadgets, everything we build, and everything we use depends on physics to work.

Even when it seems like nothing is happening at all, physics is going on all around us.

The force of gravity is pulling you down.

Muscles all over your body are also pulling on your bones, keeping you upright—even though you don't think about it.

Imagine you're sitting still on a chair, for example.

Friction makes your clothes grip the chair.

But the chair is pushing up on you. The two forces balance, so you stay still.

Physics and physicists play an important part in making things work, from everyday tools and objects to new designs and hi-tech inventions.

Making things work

When you cross a bridge, go on a flight, switch on a hairdryer, or just use a pair of scissors, it only does its job because someone has made sure that the physics works properly.

For example, a bridge has to be made of materials that are strong enough, in a shape that won't collapse.

Can you see how a suspension bridge works against gravity to hold itself up?

Four towers

Steel cables

Concrete anchors

🔍 MOTORS PAGE 42

A hairdryer works because it has a motor that turns electrical energy into heat energy.

14

Making things safe

If you're on a boat and too many people get on, it will sink. What about if you're on a fairground ride? If it moves too fast, everyone could get thrown off!

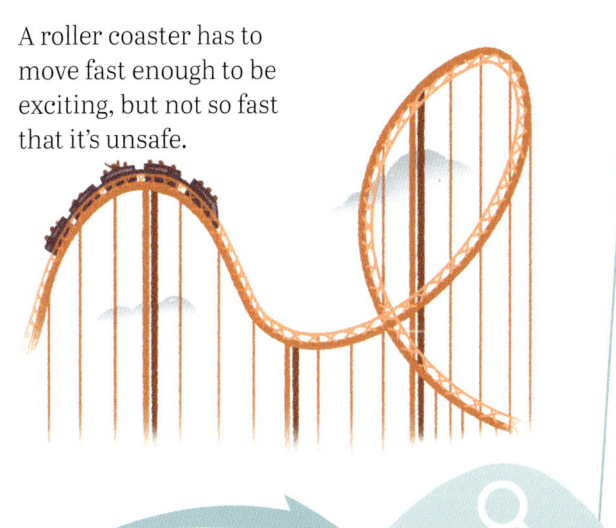

A roller coaster has to move fast enough to be exciting, but not so fast that it's unsafe.

That's where physics measurements come in. They are used to calculate the safe speeds, weights, or loads for all kinds of things—like cargo trucks and ships, elevators, swings, walkways, buses, and balconies.

🔍 MEASUREMENTS PAGE 10

Brilliant inventions

As physicists make discoveries and find out how things work, their research often leads to new inventions too—such as lasers, radio broadcasting, or solar panels.

In the 1940s, physicist Percy Spencer was working with radio waves. He noticed that they heated up a chocolate bar in his pocket—and used his discovery to invent the microwave oven.

Percy Spencer

The first microwave ovens were very big.

🔍 RADIO WAVES PAGE 38

Solving mysteries

Physics is also the science that tries to answer the biggest questions of all, such as:

- Why does matter exist?
- How can matter have gravity that can pull across empty space?
- How does time work?

We still don't really know the answers to these questions, but physicists are working on them ...

🔍 MATTER MYSTERIES PAGE 76

15

What is energy?

Energy is incredibly important in physics, but it can be hard to explain exactly what it is. Usually, it's defined as the ability to do work or make things happen. But what does that mean?

Understanding the different types of energy and how they work is a central part of physics.

Basically, energy is what makes everything happen, move, or change in any way.

A moving ball has energy, in the form of movement.

A burning flame gives out heat and light energy.

Food contains energy. Humans and other animals eat food to get energy to make their bodies work.

Energy makes everything in the Universe work. Planets and moons orbiting, galaxies spinning, and stars glowing all use energy.

Forms of energy

Energy comes in several forms, or types.

The main forms of energy are:
- Kinetic energy (movement energy)
- Potential energy
- Heat energy, also called thermal energy
- Chemical energy
- Sound energy
- Light and other types of electromagnetic energy, such as X-rays
- Electrical energy
- Nuclear energy

Different forms of energy can happen at the same time. A flying helicopter has movement (kinetic) energy, but it also makes a lot of sound energy.

SOUND ENERGY
PAGE 32

KINETIC ENERGY
PAGE 18

Energy, work, and power

Physicists often talk about work and power, as well as energy.

Work happens when one type of energy is converted or changed into another. It's measured in joules (J).

For example, if you carry two heavy bags of groceries home, you're doing work by changing energy from food into movement energy.

Power is the rate, or speed, that work happens at. It's measured in watts (W).

For example, maybe you can carry two bags home in five minutes.

But someone bigger and stronger could carry four bags in the same time.

They have more power, so have more energy and can do more work per minute.

🔍 RENEWABLE ENERGY PAGE 45

Staying the same

An important thing to know about energy is that it can't be created or destroyed. Instead, it can only change form. This is called the law of conservation of energy.

Sound energy

🔍 ELECTRICITY PAGE 40

Electrical energy in wire

For example, when you switch on a speaker, electrical energy changes into sound energy.

Energy and matter

It's also possible for energy to change into matter, and for matter to change into energy. This means that we can see matter as a form of energy, too.

Nuclear power plants and nuclear weapons work by turning a small amount of matter into a huge amount of energy.

🔍 NUCLEAR POWER PAGE 75

17

Kinetic energy

Kinetic energy means the energy of movement or motion. A moving object, like a train zooming along a track, has kinetic energy.

Everything that moves has kinetic energy, including big and small movements, big and small moving objects, and objects that move in different ways.

A tiny leaf fluttering on a tree …

A rocket taking off …

In fact, everything is moving, all the time— even if it doesn't seem to be …

An ice skater spinning …

A swing swinging to and fro …

Making movement energy

Like all energy, kinetic energy cannot be created from nothing. It can happen when another form of energy is converted into movement. Or it can happen when one type of movement causes another.

For example, chemical energy in your food gets converted into movement when you run.

CHEMICAL ENERGY PAGE 28

Ball moves

Leg moves

If you kick a ball, the movement of your leg makes the ball move.

Moving particles

All matter has kinetic energy, even when it's staying still, because the tiny atoms it's made of move, too.

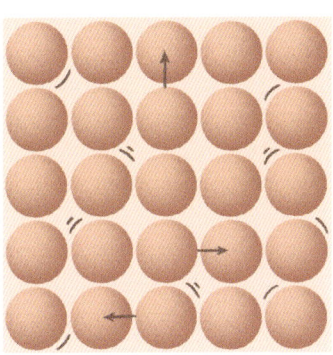

In **solid** materials, they jiggle to and fro slightly.

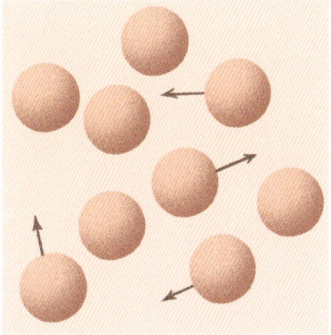

In **liquids**, they move around more freely.

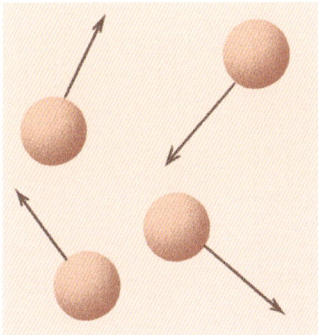

In **gases**, they zoom around at high speed.

STATES OF MATTER PAGE 70

How much kinetic energy?

The amount of kinetic energy of an object depends on two things:
- Its mass, which is the amount of matter it contains ...
- ... and how fast it's going.

MATTER PAGE 68

Imagine you threw two balls made of different materials at a glass window.

 *A hollow ping-pong ball*

 *And a solid steel ball the same size*

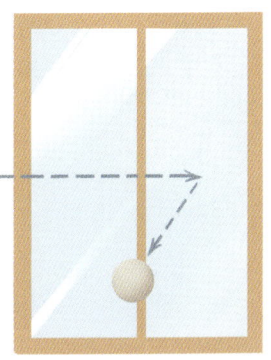

The ping-pong ball has much less mass and would have less kinetic energy.

It wouldn't have enough energy to break the glass.

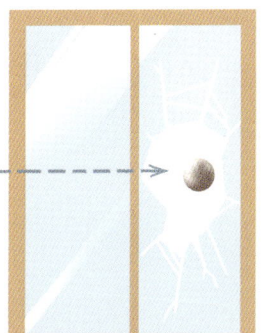

The iron ball has a lot more mass, and would take more energy to throw.

As it moved through the air, it would have more kinetic energy.

If it hit a window, it could break the glass.

Multiple movements

An object can move in different ways at the same time, too.

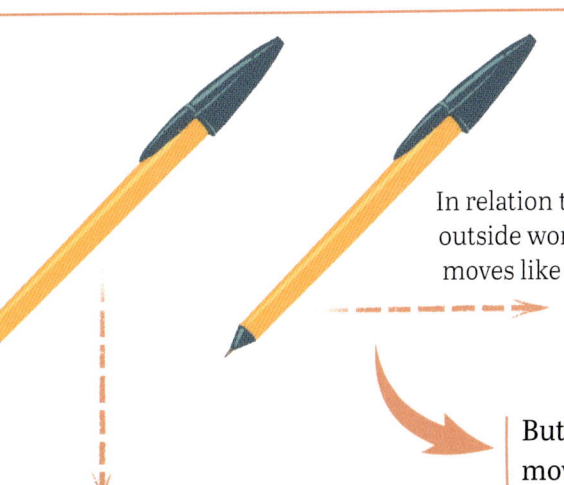

Imagine you're on a fast-moving train, sitting at a table. You pick up a pen and drop it.

In relation to the train, it falls straight down.

In relation to the outside world, it moves like this.

But the pen is actually moving forward at high speed, along with the train.

The same is true for the Earth. It's zooming through space as it orbits the Sun. So everything that moves on Earth is moving through space, too!

ORBITS PAGE 84

Potential energy

Potential energy is stored energy that is held in an object because of its position.

For example, imagine a big boulder at the top of a mountain. Because it's so high up, it has the potential to roll all the way to the bottom. When it begins to roll, its potential energy is converted into kinetic (movement) energy.

Potential energy

Kinetic energy

Potential energy can work in a variety of ways. Just like other forms of energy, it gets converted to or from another type of energy—usually movement.

Potential everywhere

Potential energy is always being stored and released all around us. The water cycle is one example.

The Water Cycle

It cools and forms clouds. The water in the clouds has potential energy ...

... until it falls as rain, and the potential energy becomes movement.

It rises into the sky in the form of a gas, water vapor.

The Sun's warmth makes water evaporate from the sea.

FALLING OBJECTS PAGE 53

Storing it up

Potential energy can be stored up when something does work. The energy used to do the work gets turned into potential energy.

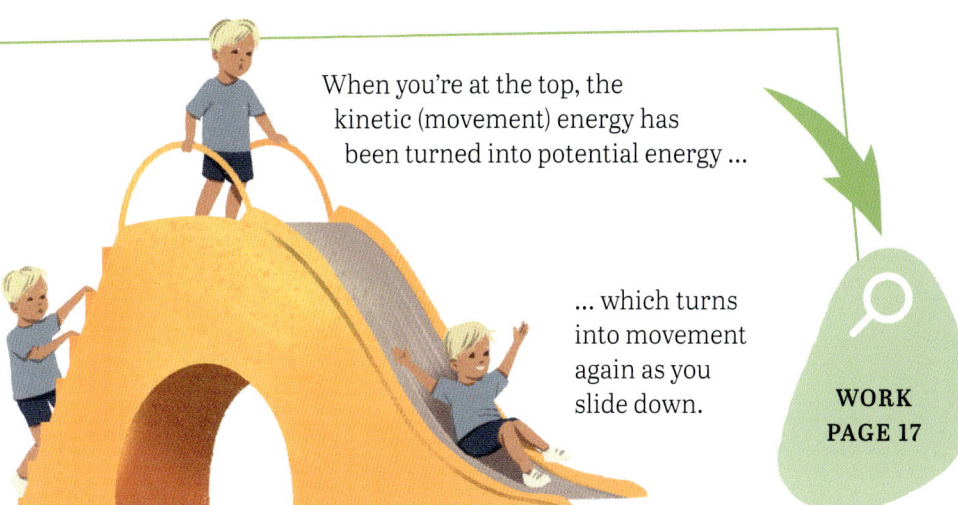

When you're at the top, the kinetic (movement) energy has been turned into potential energy ...

... which turns into movement again as you slide down.

When you climb up a slide, you do work, using energy to move your body up the steps.

 WORK PAGE 17

Stored in a spring

Another kind of potential energy gets stored when you stretch a spring or elastic band, or squash something springy like a bouncy ball.

As you stretch an elastic band, you do work and use energy.

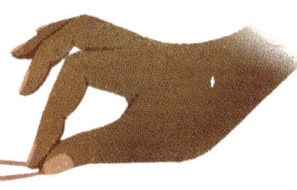

This becomes potential energy stored in the elastic band.

When you let go, the stored-up energy makes the band snap back.

TWANG!

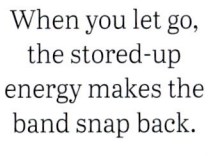
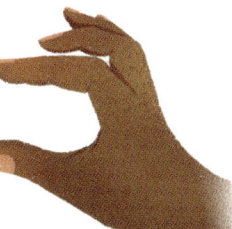

A basketball stores potential energy when it is thrown onto the ground. The movement energy makes it get squashed, storing potential energy.

Then it springs back into its normal shape, making it bounce back up.

SPRINGS PAGE 56

Many other things work this way, too:
- A bouncy diving board
- A bow and arrow
- A trampoline with springs around the edge
- Bouncy castles
- An elastic bungee jumping cord

Can you think of more?

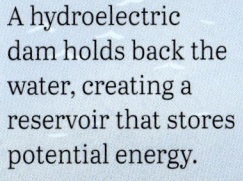

Using the potential

We can sometimes use potential energy to do work for us.

Thanks to the water cycle storing potential energy, there is always a flow of water downhill, from hills and mountains where rain falls, toward the sea.

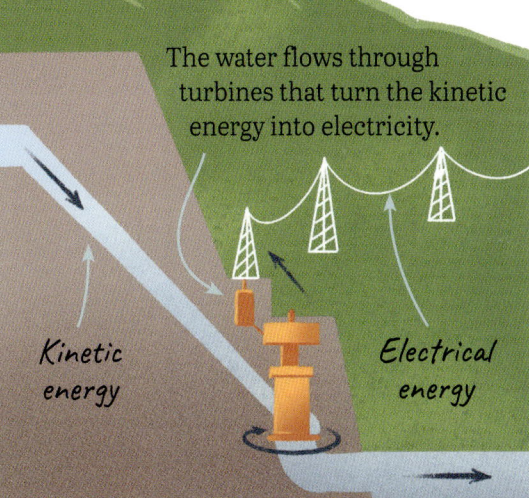

A hydroelectric dam holds back the water, creating a reservoir that stores potential energy.

The water flows through turbines that turn the kinetic energy into electricity.

Potential energy in water *Kinetic energy* *Electrical energy*

RENEWABLE ENERGY PAGE 45

Heat energy

A cup of hot chocolate, a bubble bath, or a hot water bottle all feel warm. An ice cube or a drink from a refrigerator feel chilly. But what exactly is heat?

Heat is a form of energy. Physicists also call it thermal energy. It's actually a kind of kinetic (movement) energy.

The tiny atoms and molecules that matter is made of move all the time. The faster they move, the more heat energy they have.

When we talk about how "hot" something is, we really mean how much kinetic energy its atoms have—in other words, how fast they are moving and bumping into each other.

Cold drink

Even when an object feels cold, its atoms are moving.

Hot drink

When the cup is hotter, its molecules move faster and have more energy.

Everything has heat!

All matter, even if it feels really cold to us, has some heat energy. That's because the atoms and molecules in matter are always moving. Some move less than others, but any movement counts as heat energy.

So even an icicle in the Antarctic or a bag of frozen peas have some heat energy. They just have less than warmer things, such as a cat.

Some heat

More heat

🔍 TYPES OF ATOMS PAGE 68

Sensing heat

If we we think of something as hot or cold, that really just means it's more or less hot than we are.

If something has less heat than your skin, it will feel cool or cold. That's because you can sense it cooling your skin. Hotter things heat up your skin and feel warm.

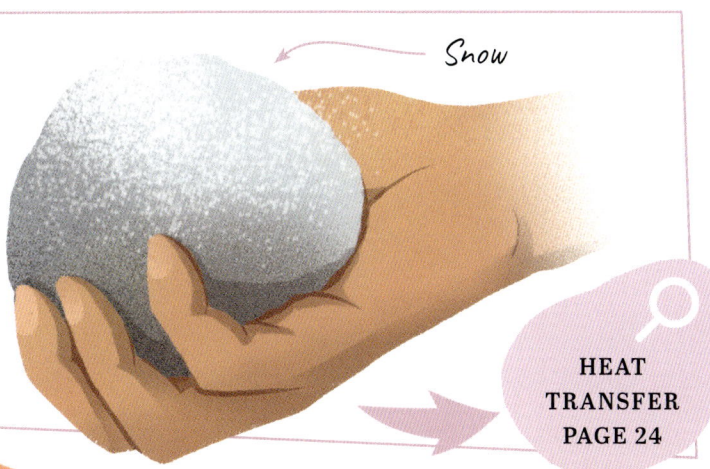
Snow

🔍 HEAT TRANSFER PAGE 24

Temperature

The temperature of matter means how hot or cold it is. There are three different scales for measuring temperature: Fahrenheit, Celsius, and Kelvin.

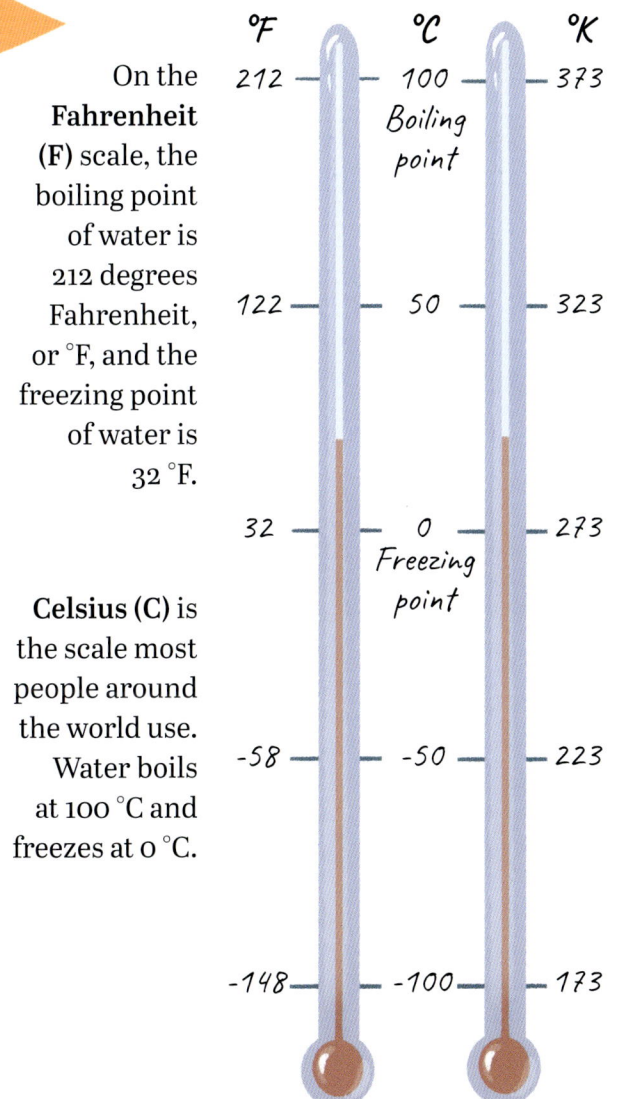

On the **Fahrenheit (F)** scale, the boiling point of water is 212 degrees Fahrenheit, or °F, and the freezing point of water is 32 °F.

Kelvins (K) are like Celsius degrees, but the Kelvin scale starts at a much lower temperature. Water boils at 373 °C and freezes at 273 °C.

Celsius (C) is the scale most people around the world use. Water boils at 100 °C and freezes at 0 °C.

Absolute zero

The Kelvin (K) temperature scale starts at a very low temperature known as "absolute zero." In Celsius, it's 273.15 degrees below freezing. This is the temperature at which physicists think the atoms in matter would stop moving completely.

Atoms

0 K (−273.15 °C) — No movement

However, they have not managed to cool anything down to this temperature—almost, but not quite.

🔍 WATER PAGE 69

🔍 FREEZING PAGE 71

Heat transfer

Heat transfer means the way heat moves or spreads from one place or object to another. For example, if you put your hands near a hot radiator, they'll warm up, too. But how?

Heat always spreads from warmer objects or substances to cooler ones, until they balance out and are the same temperature.

Cooler air

Hot toast

All the same temperature

Heat energy, or thermal energy, can be transferred in three different ways, known as conduction, radiation, and convection.

Think of a piece of hot toast. As soon as you take it out of the toaster, it starts cooling down. Its heat spreads to the plate and the surrounding air. Eventually, they are all the same temperature.

Conduction

Conduction happens when heat spreads through an object or between objects that are touching.

When things are hotter, their atoms have more energy and move more. They push on the atoms next to them, so they move more, too, and warm up.

Heat spreading through an object.

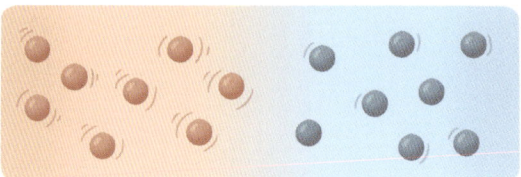

Heat energy spreads from one atom to the next.

Heat spreading between touching objects.

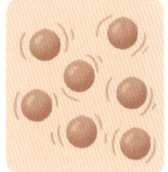
Warmer object

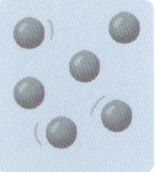
Cooler object

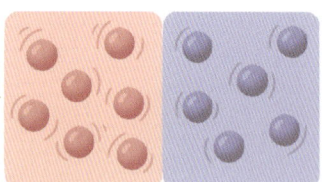
Heat energy in one object spreads to atoms in the other object.

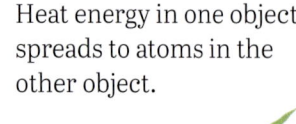

As an example, imagine putting a teaspoon into hot tea.

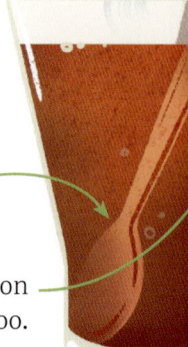

The heat energy from the tea spreads into the spoon.

It also spreads up the spoon handle, so that gets hot, too.

TYPES OF ATOMS PAGE 68

Radiation

You can feel the heat of a campfire, even if the air in between is cold. The Sun can warm you up too, even though it's 150 million km (93 million miles) away across empty space. This is because of radiation.

Everything that has heat energy gives out energy waves called infrared radiation.

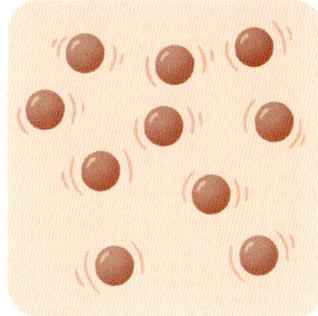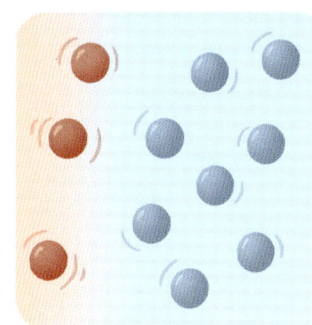

Infrared waves

When the infrared waves hit another object, they make its atoms and molecules move faster, and it warms up.

🔍 **INVISIBLE LIGHT WAVES PAGE 38-39**

Convection

Convection makes heat spread through a substance as warmer atoms and molecules move around. It happens most in liquids and gases.

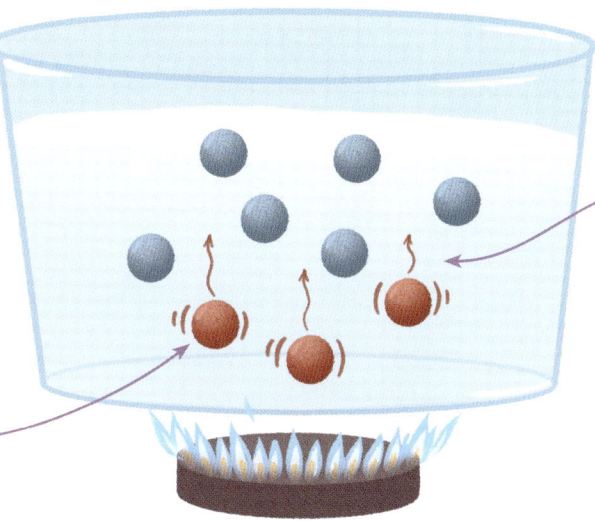

🔍 **STATES OF MATTER PAGE 70**

The warmer water rises up, and cooler water sinks to take its place.

As a substance such as water heats up, some atoms get heated first.

They move faster and spread farther apart, making the warmer water lighter.

In this way, heat eventually spreads through all the water.

Conductors and insulators

Some materials carry, or conduct, heat much more easily and quickly than others.

A steel iron warms up quickly and gets very hot.

🔍 **MATTER PAGE 68**

For example, steel is a good conductor of heat.

Plastic is a poor conductor of heat, also called an insulator. Heat does not spread through it easily, so it's used to make the handle, which stays cool.

25

Heat expansion

When materials heat up, they expand, or get bigger. This is known as heat expansion or thermal expansion.

As a material heats up, its atoms move more. This makes them push away from each other and get farther apart, and the material expands.

In a solid, the atoms or molecules stay fixed in place, but move slightly farther apart.

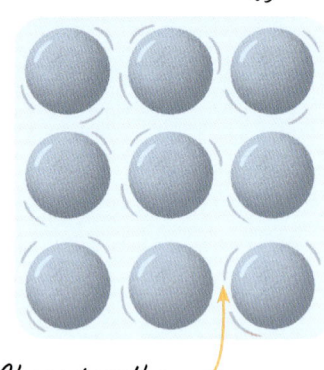

Less heat energy — Closer together

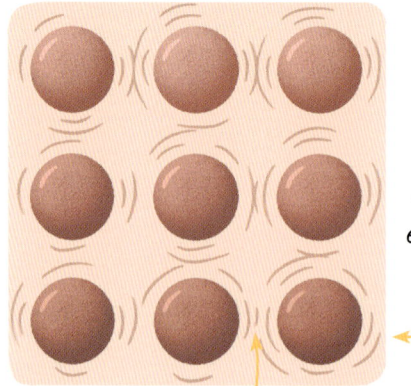

More heat energy — Farther apart — Object expands

Heat expansion has all kinds of effects and uses in everyday life.

In liquids and gases, the atoms or molecules move around freely. As they heat up and move faster, they can spread much farther apart, and the material expands more.

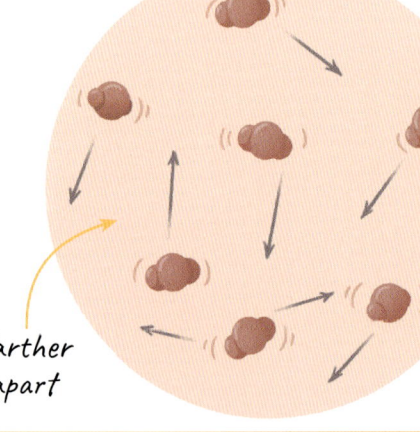

Less heat energy — Closer together — More heat energy — Farther apart

MAKING MOLECULES PAGE 69

HEAT PAGE 22

Space to grow

When engineers and architects design buildings, bridges, and railroads, they have to allow for heat expansion.

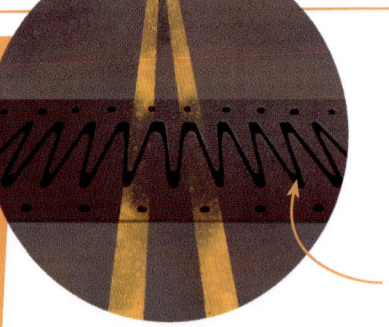

Gap

Thanks to heat expansion, skyscrapers and towers grow taller in hot weather.

The Eiffel Tower in Paris can grow up to 15 cm (6 in) taller in the summer.

For example, bridges are often built in sections with expansion joints between them. These joints have a gap or a rubbery material that can be squashed, so that the bridge sections can expand safety in hot temperatures.

Expansion joint

26

Thermometer expansion

Thermometers are used to measure temperature. Traditional thermometers work using heat expansion.

The thermometer contains a liquid, which is usually red to make it easy to see. The hotter it gets, the more the liquid expands, making it push farther up inside the tube.

A warm summer temperature, 28°C or 82°F

A cold winter temperature, 4°C or 39°F

🔍 **TEMPERATURE PAGE 23**

Rising seas

Heat expansion affects the seas and oceans, too. As water gets warmer, it takes up more space.

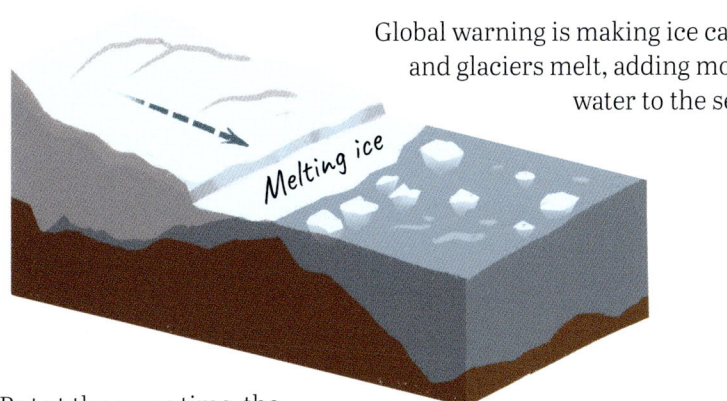

Global warning is making ice caps and glaciers melt, adding more water to the sea.

Melting ice

But at the same time, the seas and oceans are warming up and taking up more space—making the sea level even higher.

Warmer water means a higher sea level.

Cooler water means a lower sea level.

Bigger and lighter

Hot-air balloons use heat expansion to work.

When the burner heats the air inside the balloon, it expands.

As the atoms and molecules are farther apart, the air in the balloon is less dense (heavy for its size) than the cooler air around it.

The cooler, heavier air sinks, and the lighter balloon floats upward.

🔍 **LIQUIDS PAGE 70**

🔍 **FLOATING PAGE 59**

27

Chemical energy

Chemical energy is energy stored in the chemicals in a substance or material. For example, wood contains chemical energy. If a piece of wood burns, the chemical energy is released, and turns into heat energy.

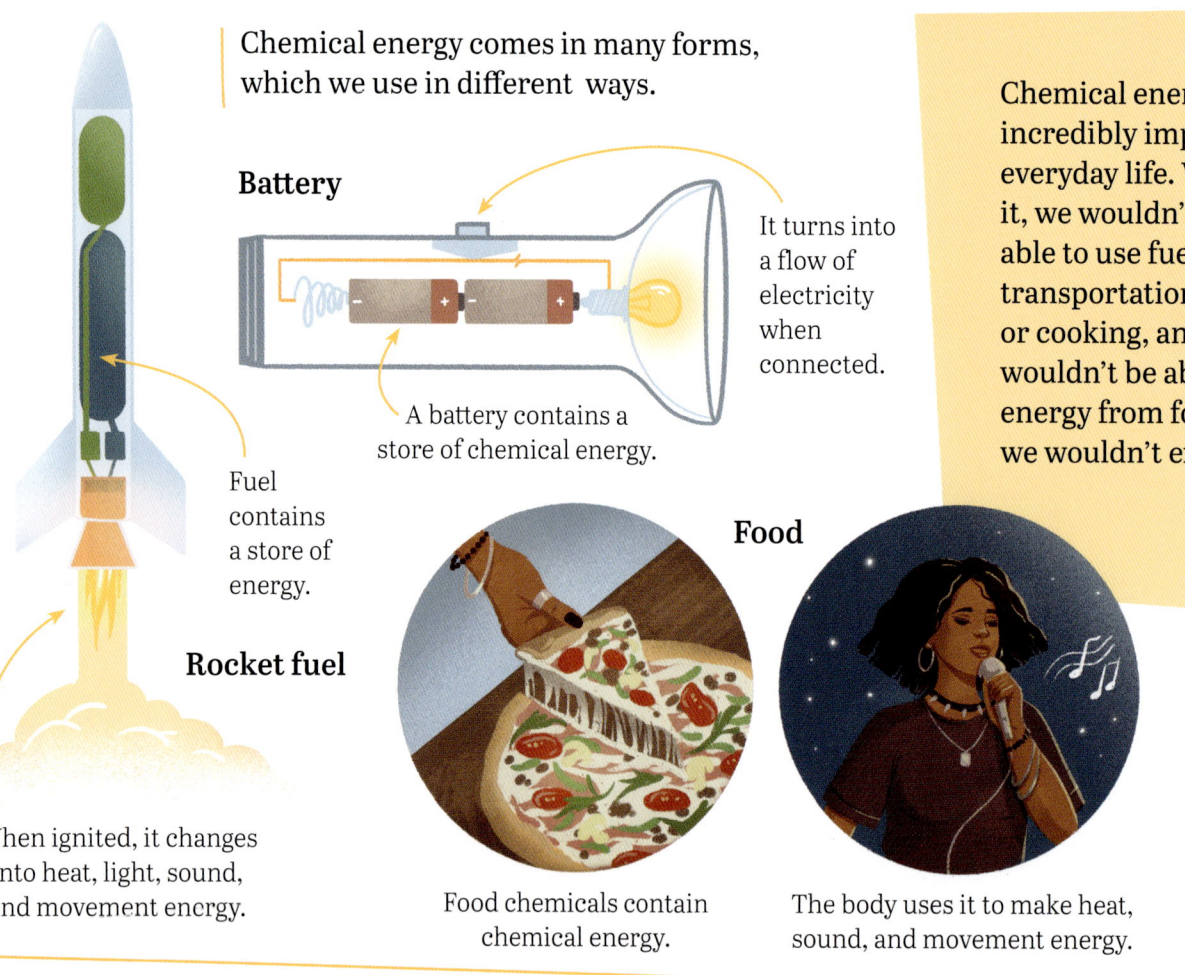

Chemical energy comes in many forms, which we use in different ways.

Battery

A battery contains a store of chemical energy.

It turns into a flow of electricity when connected.

Rocket fuel

Fuel contains a store of energy.

When ignited, it changes into heat, light, sound, and movement energy.

Food

Food chemicals contain chemical energy.

The body uses it to make heat, sound, and movement energy.

Chemical energy is incredibly important in everyday life. Without it, we wouldn't be able to use fuel for transportation, heating, or cooking, and we wouldn't be able to get energy from food—so we wouldn't exist!

How it works

Chemical energy is stored in chemical bonds—the links between atoms that hold them together to make molecules. When a molecule breaks apart during a chemical reaction, energy can be released.

🔍 **MAKING MOLECULES PAGE 69**

This is a sugar molecule. In the body, oxygen and sugar react together. This forms new chemicals and releases energy for the body to use.

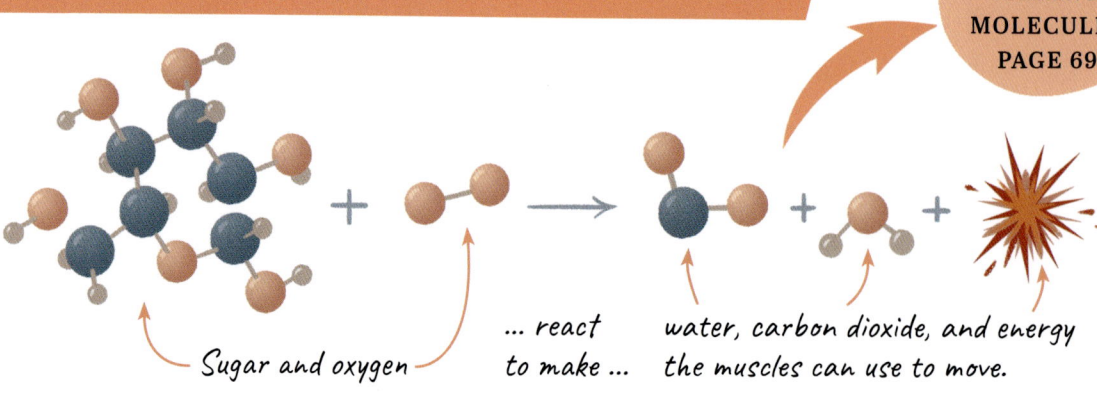

Sugar and oxygen ... react to make ... water, carbon dioxide, and energy the muscles can use to move.

28

Food calories

You've probably heard of calories in food. Calories are a measurement of how much energy food contains.

For example, sugar contains a lot of energy: about 387 calories per 100 g.

Different foods can contain very different amounts of energy.

Food	Calories per 100 g
Cheese	402
Broccoli	34
Milk chocolate	535
Lentils (cooked)	125
Eggs	155
Bananas	89

UNITS OF MEASUREMENT PAGE 11

The power of plants

Plants make food by soaking up sunlight and converting it into chemical energy stored in their leaves, fruits, seeds, and other parts—a process called photosynthesis.

In turn, this chemical energy becomes food for plant-eating animals, and they become food for meat-eating animals.

So plants provide the chemical energy that supports most living things.

LIGHT PAGE 34

Bang!

Some substances have a lot of chemical energy that is released very quickly when they burn, causing an explosion.

Fireworks contain a powder made of a mixture of chemicals, along with smaller amounts of materials that burn in different colors.

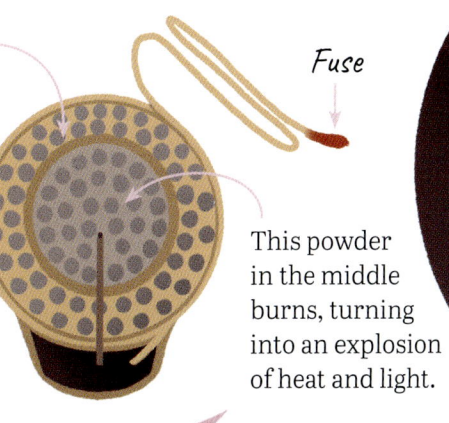

This powder explodes and makes the firework shoot upward.

Fuse

This powder in the middle burns, turning into an explosion of heat and light.

HEAT ENERGY PAGE 22

29

Energy waves

Waves are vibrations that can carry energy from one place to another. Some waves can travel through matter, making it vibrate, but they don't carry matter with them—just energy.

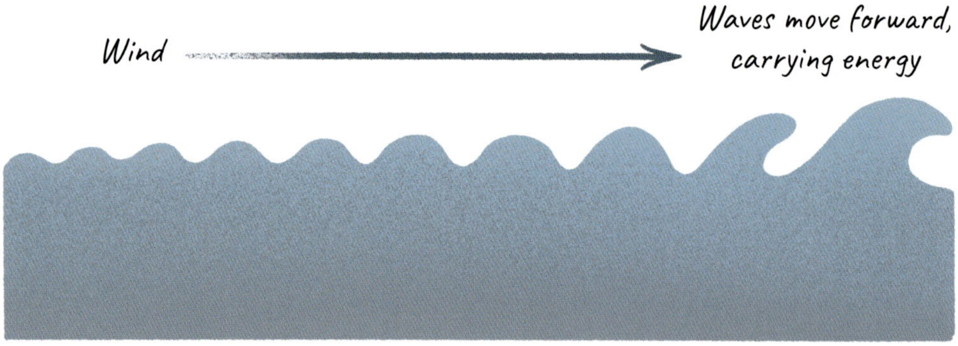

Wind → *Waves move forward, carrying energy*

Waves in water are one example. In the sea, the wind makes waves that travel across the surface of the sea.

But the waves don't make the water travel with them. As a wave passes by, the water just moves up and down. If you were in a boat on the sea, you'd bob up and down as each wave went by.

Boat moves up and down

Waves move forward

Several different types of energy can travel in the form of waves, including kinetic (or movement) energy, sound, and light.

Waves in matter

Waves on the sea or ripples on a pond travel through water, which is a type of matter. If there was no matter there, the waves couldn't exist.

Sound also travels in waves that move through matter. Instead of making it vibrate up and down, sound waves make molecules in the air vibrate forward and backward very fast.

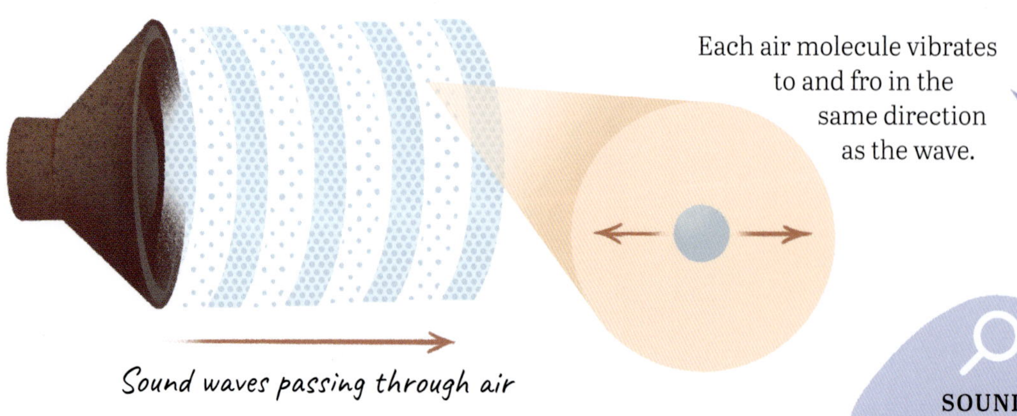

Sound waves passing through air

Each air molecule vibrates to and fro in the same direction as the wave.

🔍 SOUND PAGE 32

Waves in empty space

Electromagnetic energy waves are a different type of wave that can travel through empty space, or a vacuum. They can often travel through matter, too. They include light waves, X-rays, and radio waves.

Magnetic vibrations
Electrical vibrations
λ *Wavelength*

ELECTROMAGNETIC SPECTRUM PAGE 38

It's hard to understand electromagnetic waves, but they are usually shown in diagrams like this. The wave is made up of two kinds of vibrations: electrical and magnetic. They both travel forward, like waves in water, but at right angles to each other.

Long-distance travel

Waves can carry energy a long way. For example, ocean waves can travel thousands of kilometers or miles before crashing onto a shore.

Journey of an ocean wave

Light waves from a star

STARS PAGE 82

Electromagnetic waves, such as light waves, can travel vast distances across outer space. That's how we can see stars that are billions or trillions of kilometers away.

Parts of a wave

Physicists have names and measurements for the different parts of a wave:

MEASUREMENTS PAGE 11

The **peak** is the highest point of the wave.

Wavelength is the distance from a point on wave to the same point on the next wave.

The **amplitude** is the strength or size of a wave.

Frequency is a measure of how many waves pass by in a particular time period, such as a minute or second.

Sea level

Wave height

The **trough** is the lowest point.

Sound

Sound is a type of energy. It's made up of vibrations that can spread through the air, or other materials, in the form of sound waves.

Here's an example—a quacking duck. To make a sound, the duck makes parts in its throat vibrate, or shake quickly to and fro.

Many animals, including humans, can hear sounds with their ears.

As the duck's throat vibrates, it pushes against molecules in the air, making them vibrate, too.

QUACK!

Each molecule pushes against other molecules.

The waves of vibrations spread out in all directions.

Sounds like a duck!

All sounds are made up of different patterns of sound waves. Sound waves are not the same as water waves. They work in a different way and are called longitudinal waves.

The vibrating air hits the eardrum inside the ear and makes it vibrate.

Inside the ear, the vibrations are turned into signals and sent to the brain.

The brain figures out what the sounds mean.

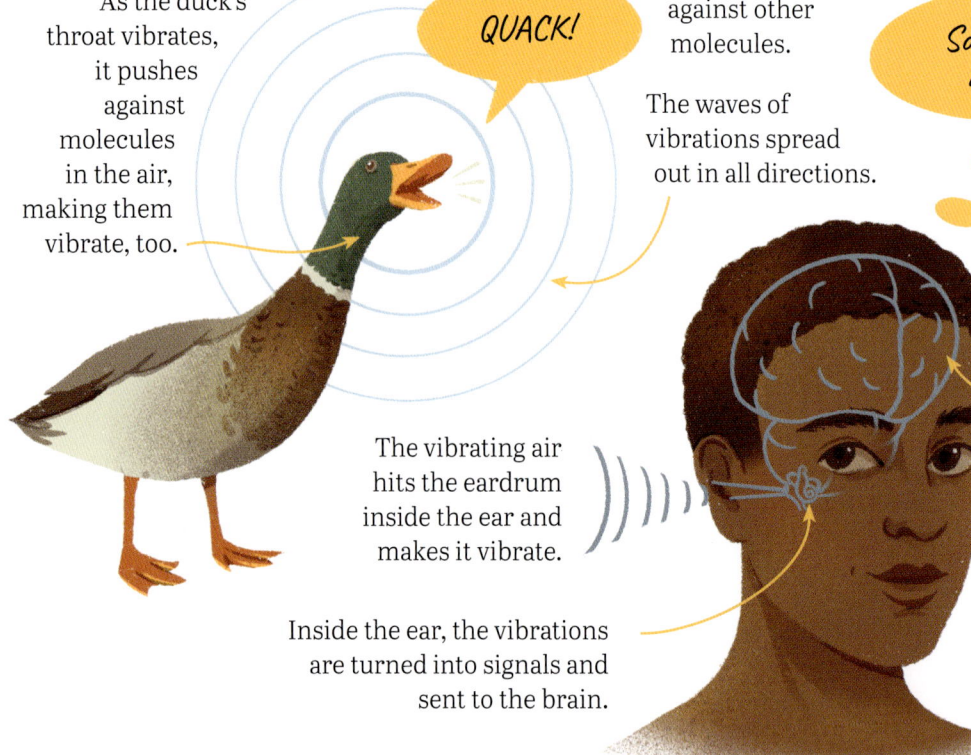

ENERGY WAVES PAGE 30

Sound waves

Sound waves spread through the air as air molecules push against each other. Instead of moving up and down like a water wave, each molecule moves forward and backward as the wave passes.

Sound waves can travel through other materials, too, like water, glass, and metal. But they cannot spread through a totally empty space, or vacuum. They need matter to travel through.

Longitudinal sound waves

Molecule moves forward and backward with each wave.

Wave direction

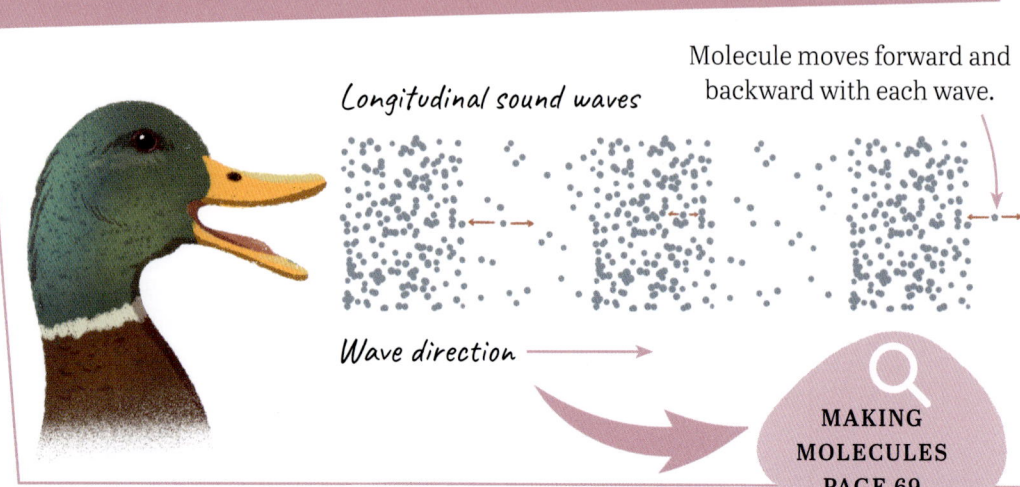

Most of space is a vacuum, so sound waves can't travel through it.

MAKING MOLECULES PAGE 69

SPACE PAGE 80

32

Making sounds

All the sounds we can hear are just patterns of sound waves made by different objects.

When you can hear lots of sounds at once, lots of different sound wave patterns are entering your ears. Your brain untangles them and identifies the separate sounds.

How loud?

The louder a sound is, the more energy it has, and the more the molecules in the waves move. Scientists use a scale called the decibel scale to measure the loudness or volume of sound.

Musical instruments

Objects falling, breaking, rolling, or bumping together.

Vehicle engines

People talking, laughing, or singing.

ENERGY WAVES PAGE 30

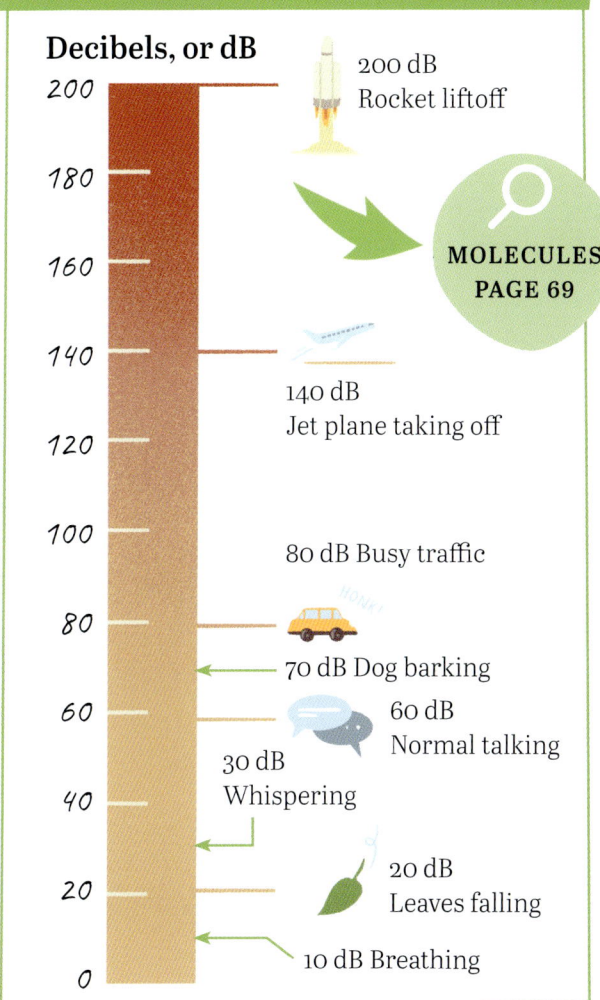

Decibels, or dB

200 dB Rocket liftoff

MOLECULES PAGE 69

140 dB Jet plane taking off

80 dB Busy traffic

70 dB Dog barking

60 dB Normal talking

30 dB Whispering

20 dB Leaves falling

10 dB Breathing

Studying sound

The study of how sound waves behave is called acoustics. Sound waves can reflect off some hard surfaces, making an echo. Other materials, like soft curtains, soak up and muffle sound energy.

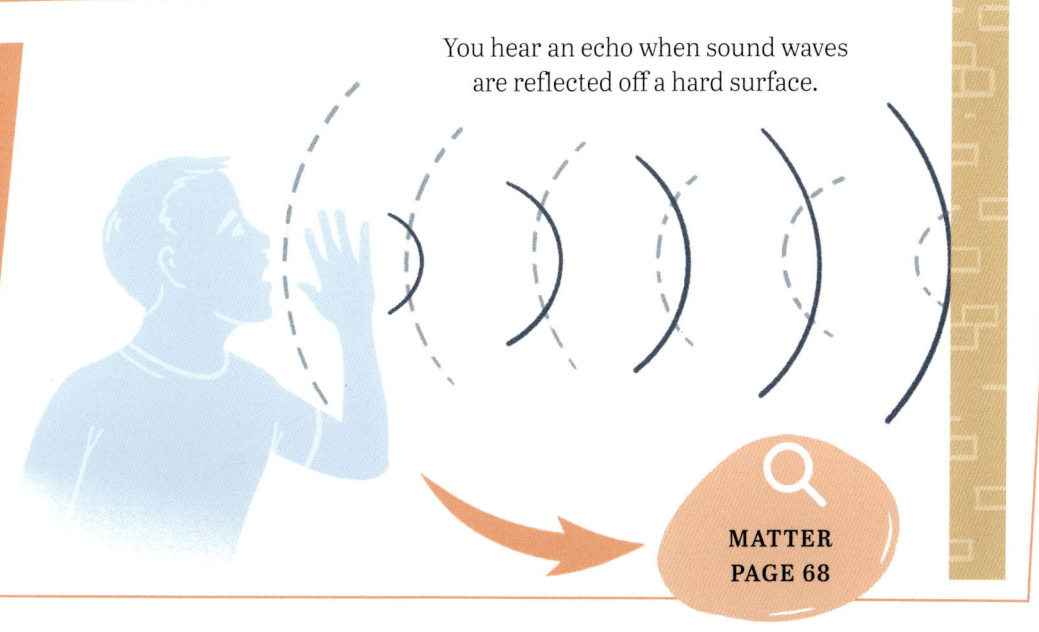

You hear an echo when sound waves are reflected off a hard surface.

MATTER PAGE 68

33

Light

If you can see something, that's because light is coming from somewhere and entering your eyes. Light is a type of energy. It can travel very fast across huge distances, in the form of light waves. That's why we can see the faraway stars!

Light sources

Light energy comes from light sources. What makes these things **glow** with light? There are several ways it can happen.

Candles and fires glow with light as fuel burns.

Some animals glow by mixing chemicals inside their body.

Light bulbs glow by turning electrical energy into light energy.

🔍 **ELECTRICITY PAGE 40**

Some materials glow with light when they get very hot or burn. Stars, including our Sun, are made of gases under huge pressure, releasing energy as heat and light.

The speed of light

Light is FAST—very, very fast. In fact, it's the fastest thing there is!

In the empty vacuum of space, it zooms along at 300,000 km per second (186,000 miles per second). That's the same as 1,080,000,000 km per hour (671,000,000 miles per hour).

- The distance light travels in a year is called a light-year.
- Light-years are used to measure distances in outer space.

🔍 **LIGHT-YEARS PAGE 81**

Light is more than a million times faster than a speeding jumbo jet!

34

Light waves

Light travels as a type of wave called an electromagnetic wave. The waves are not really like waves in water, but they look similar in diagrams.

Light waves have an electric part and a magnetic part.

Electric

Magnetic

Invisible waves

There's a whole range of other electromagnetic waves too, including X-rays and radio waves. Light waves are just the type we can see!

ELECTRO-MAGNETIC SPECTRUM PAGE 38

Bouncing everywhere

We see light directly from light sources such as stars or light bulbs. We see other objects when light bounces or reflects off them.

Light

Object

We see the reflected light coming from objects.

Sight

REFLECTION PAGE 36

The spectrum

Light comes in a range of different wavelengths. This "length" is the distance from one wave to the next. We see the different wavelengths as a range, or spectrum, of different shades.

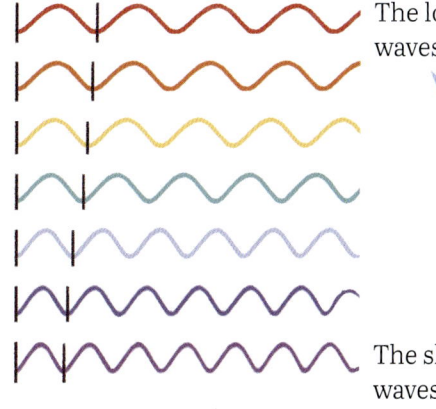

The longest waves are red.

The shortest waves are violet.

Seeing a rainbow

If this pattern looks familiar, that's because it's what you see in a rainbow! Rainbows happen when white sunlight shines into raindrops in the sky and bounces back out. As the light enters and leaves the raindrop, it bends, or refracts, and this separates it out into a spectrum of different wavelengths.

WAVES PAGE 31

REFRACTION PAGE 37

Reflection and refraction

Although light is made up of energy waves, the waves themselves travel in straight lines.

You can see how light travels in straight lines when you see rays of sunlight or light beams shining from a torch or flashlight at night.

Sunlight shines through clouds in straight rays or "sunbeams," like this.

However, rays of light can change direction. One way this can happen is reflection, when light rays bounce off a surface, such as a mirror. Another way is refraction, when light rays bend as they pass through a transparent material.

Shadows also happen because light travels in straight lines. When an object blocks light rays, they can't bend around it, so the object casts a shadow.

Shadow is the same shape as object

Light rays — Object — Shadow

MATTER PAGE 68

Reflection

When a ray of light hits a smooth, shiny surface like a mirror, it will reflect, or bounce off.

All materials reflect some of the light that hits them.

If a light ray hits a mirror at an angle, it will bounce off at the same angle and change direction.

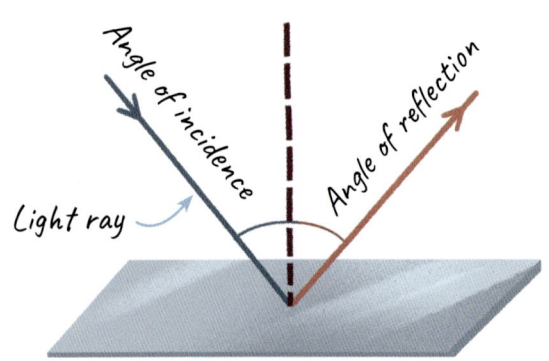

Angle of incidence — Angle of reflection — Light ray — Mirror

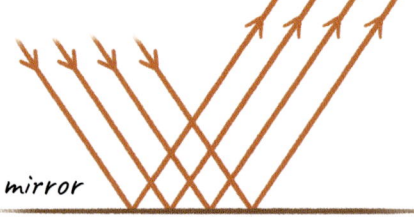

Smooth mirror

Bumpy surface

A smooth, shiny material reflects all the light rays in the same way. That's why you can see a clear reflection in a mirror.

A bumpy or matte material reflects light in different directions, so you don't see a clear reflection.

Seeing colors

Light has a range, or spectrum, of different wavelengths, which we see as different shades. Different objects reflect different wavelengths or mixtures of wavelengths.

For example, if something looks green, that's because it absorbs the other wavelengths and reflects the green wavelength.

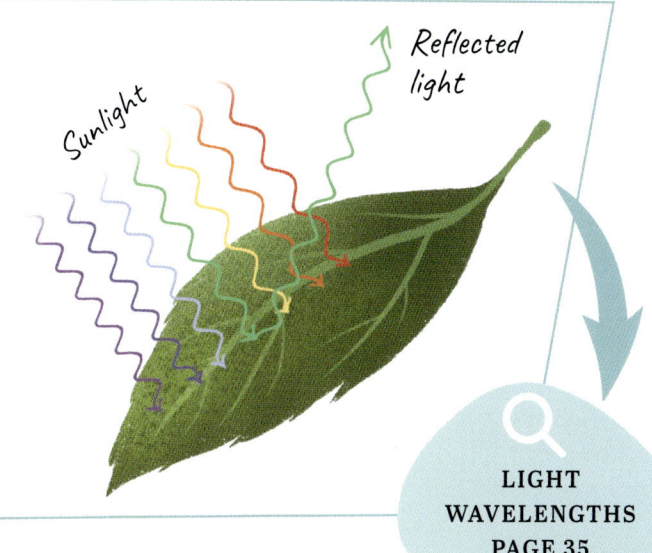

LIGHT WAVELENGTHS PAGE 35

Refraction

Refraction can happen when a ray of light moves from one transparent substance into another. The light ray changes speed, and this makes it change direction.

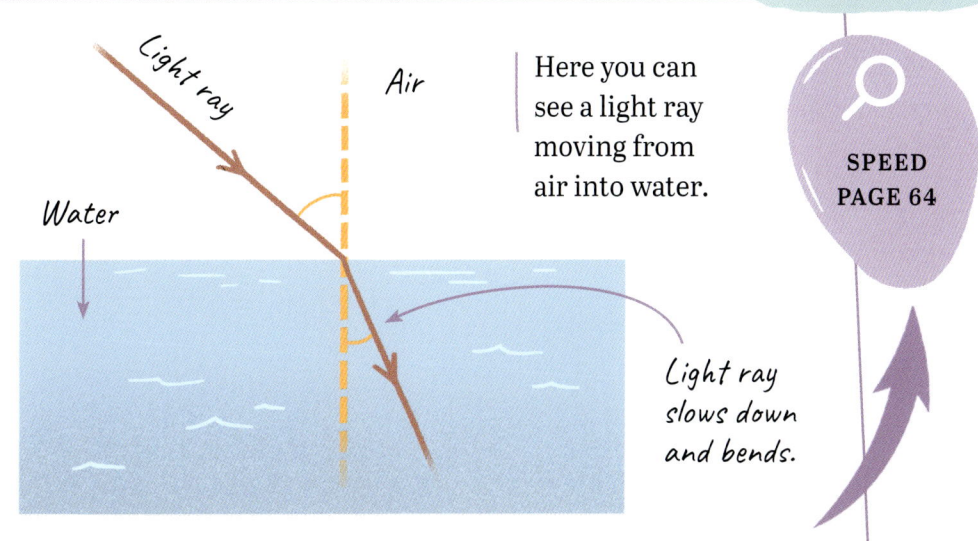

Here you can see a light ray moving from air into water.

Light ray slows down and bends.

SPEED PAGE 64

Light inventions

We use reflection and refraction in many inventions and everyday objects.

Fiber-optic cable

Reflection:
- Mirrors.
- Road studs reflect car headlights to light up road markings.
- Fiber-optic cables carry light along inside a reflective tube.

Refraction
- Lenses bend light to make magnifying glasses, telescopes, and eyeglasses work.
- Rainbow crystals refract and bend sunlight, separating it into different wavelengths and colors.

Crystal refracts and bends light, splitting it into different wavelengths and making a rainbow.

Ray of sunlight

RAINBOWS PAGE 35

The electromagnetic spectrum

Electromagnetic waves have a range of different wavelengths, known as the electromagnetic spectrum, or EMS. It's often shown as a strip-shaped diagram, like this.

Radio waves

Microwaves

Light waves are near the middle. They are the only EM waves that humans can see.

Light waves

At one end, the waves are longer, and carry less energy.

This wave shows the different wavelengths at different parts of the spectrum.

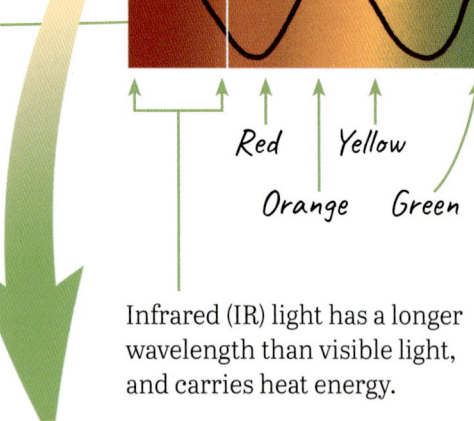

Red, Orange, Yellow, Green

Infrared (IR) light has a longer wavelength than visible light, and carries heat energy.

Radio waves

Radio waves are the longest EM waves, with wavelengths from a few centimeters to many kilometers. By modulating radio waves or adding patterns to them, we can use them to send information over long distances.

Radio waves are used in lots of inventions:
- Radio and TV broadcasting
- Remote-controlled toys and drones
- Cell phones
- Radar, used for detecting aircraft
- Radio telescopes
- Microwave ovens, which use shorter radio waves known as microwaves

Visible and invisible light waves

Visible light waves include the spectrum of wavelengths that we see as the rainbow, from longer-wavelength red light to shorter-wavelength violet light.

RADIO TELESCOPES PAGE 87

HEAT ENERGY PAGE 22

Electromagnetic, or EM waves, are energy waves that can travel across empty space. Light waves are one type of EM wave, but there are others, too.

ATOMS
PAGE 68

X-rays

Gamma rays

At the other end, the waves are very short, and carry the most energy.

Electromagnetic waves are emitted (sent out) by atoms, when parts inside them change position. The wavelength of an EM wave depends on the type of atom and the amount of energy.

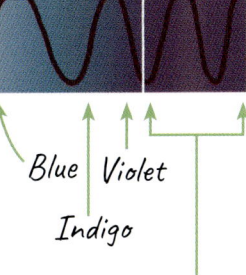

Blue / Violet
Indigo

Ultraviolet (UV) light has a shorter wavelength than visible light, and more energy. It is part of sunlight and can cause sunburn.

There are also two types of light that are invisible to humans: infrared (IR) and ultraviolet (UV).

X-rays

X-rays are short-wavelength EM waves with high energy. They can pass through the human body and can be harmful. But they can be used in small amounts to make X-ray images of our insides.

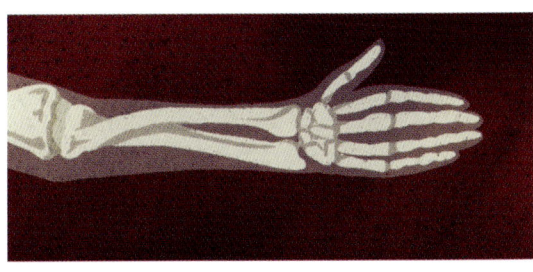

This X-ray image shows arm bones.

Gamma rays

Gamma rays are very high-energy EM waves with very short wavelengths. They are emitted by very hot objects or radioactive atoms. They are harmful to living things. However, they can be helpful too, since they are used to kill cancer cells.

An atomic nucleus (the part in the middle of an atom) emits a gamma ray.

The wavelength of a gamma ray can be smaller than the width of a single atom.

IONIZING RADIATION PAGE 72

WAVELENGTHS PAGE 31

What is electricity?

Electricity is a form of energy. It's very important to humans, since we use it to make all kinds of things work, such as lighting, heating, stoves, vehicles, TVs, phones, and computers.

Electricity happens when tiny particles move through a material. To understand how it works, you have to start with atoms, the building blocks that matter is made of.

There are many different types and sizes of atoms, but they all have the same main parts: protons, neutrons, and electrons.

Proton

Neutron

Electron

Electrons are much smaller and whirl around the nucleus.

Protons and neutrons form the central part, or nucleus.

The charge on electrons is what makes electricity work. It can make electrons break free from their atoms and flow through a material, carrying energy with them.

Protons and electrons have a quality called "charge."

Protons have a positive charge, shown as +.

Electrons have a negative charge, shown as −.

Neutrons don't have any charge.

ATOMS PAGE 68

Moving electrons

If an atom or an object has the same number of protons and electrons, they balance each other out. But when they are unequal, they create an electric charge.

This makes electrons try to move. If they can, they will flow or jump from one place to another—and that movement is electricity.

A positive electric charge pulls electrons toward it.

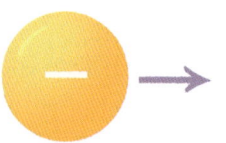

A negative electric charge pushes electrons away.

ELECTRONS PAGE 68

Current electricity

Current electricity is electricity that is flowing through a material, such as a metal wire.

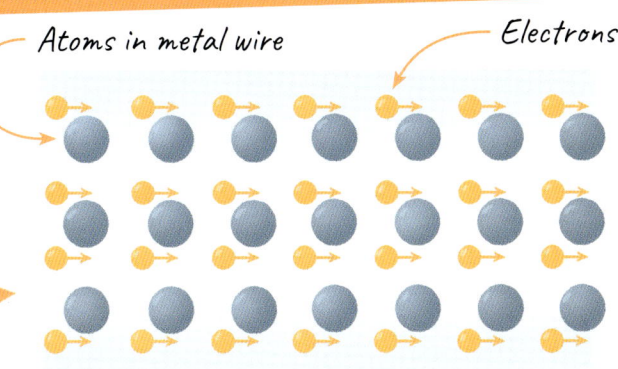

Atoms in metal wire — Electrons

The electrons move through the metal by passing between the atoms.

Electrons can only do this in materials that let them move easily, known as electrical conductors. They include metals such as copper, silver, gold, and iron.

Wires used to carry electricity are often made of copper.

WIRES PAGE 47

Static electricity

Static electricity means "still" electricity. Instead of flowing, electrons collect in one place. This can happen in materials that do not conduct electricity well, called insulators.

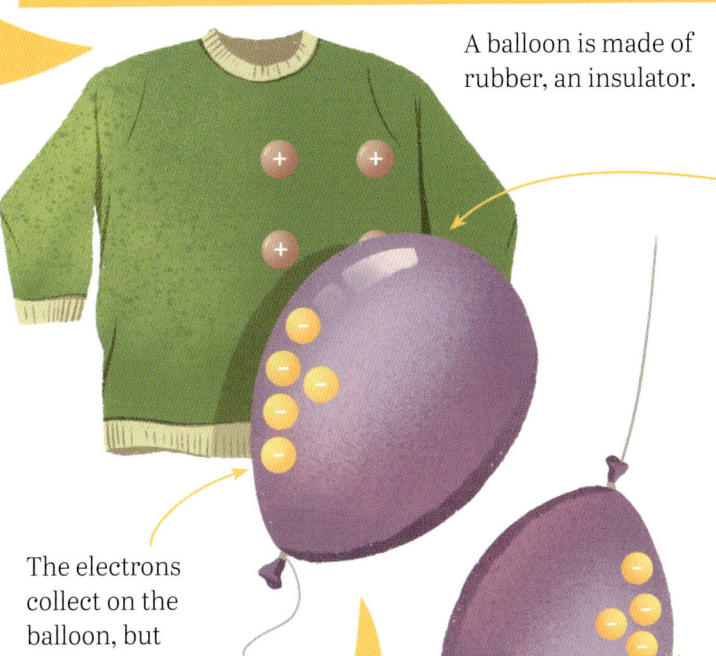

A balloon is made of rubber, an insulator.

If you rub it on a wool sweater, it collects extra electrons from the sweater.

The electrons collect on the balloon, but cannot flow away, giving it a negative electric charge.

If you do this to two balloons, they will both have a negative charge and will push each other away.

RUBBER PAGE 56

Using electricity

Electricity is very useful. It can easily be carried along wires and turned into other forms of energy, such as motion, light, heat, and sound. Most people have an electricity supply in their homes, and use lots of electric gadgets, machines, and computers.

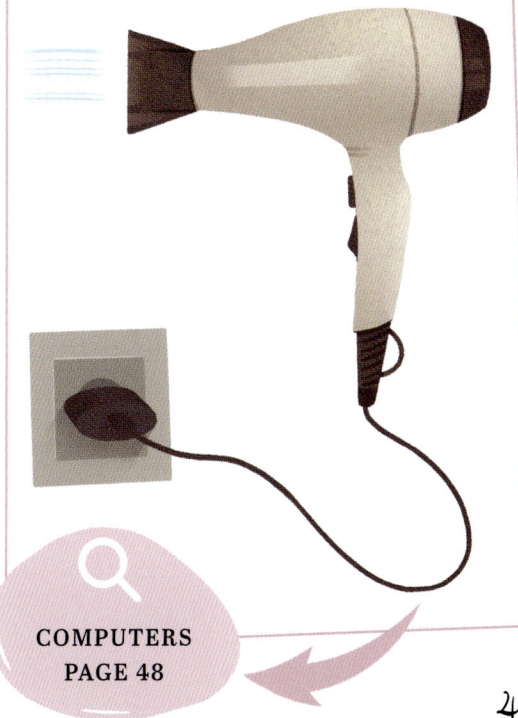

COMPUTERS PAGE 48

41

Electric circuits

Electricity can only flow if there is a complete, unbroken loop of wire, or another conductor, for it to flow around. This loop is called an electric circuit.

Here's a picture of a simple electric circuit. The two ends of the wire are linked to the battery, which provides the electrical charge that makes electricity flow.

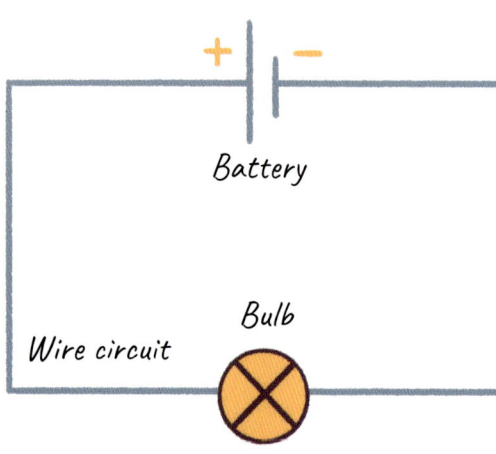

As electricity flows through the bulb, it's converted to light energy as the bulb lights up.

All circuits need a power source and a loop of conducting material. They can also include many other parts, such as bulbs, buzzers, and motors. Circuits can be simple or very complicated, and there are many different types.

Physicists show electric circuits as diagrams. This is a diagram of the circuit above.

Power sources

The power source in a circuit, such as a battery, has two connecting points or terminals. One has a positive charge and one has a negative charge. When the ends of a wire are linked to the two terminals, electricity flows around the circuit.

There are other types of power sources, too. For example, an electric socket in the wall is a power source and has positive and negative terminals.

The electric cord or cable contains two separate wires.

Electricity flows along one wire, through the bulb and then back along the other wire to the other terminal, forming a circuit.

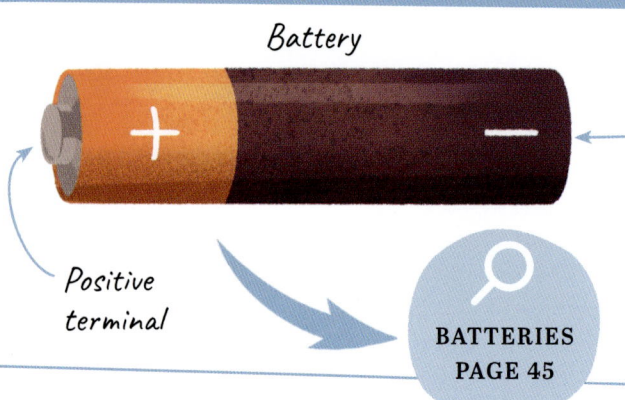

BATTERIES PAGE 45

42

Circuit parts

Circuits also have other parts, including a "load"—the thing that switches on or works when the electricity flows though it. In the diagrams on this page, it's a light bulb, shown in circuit diagrams like this: ⊗

Other circuit parts include:
- Switch—switches off the current by making a gap in the circuit.
- Buzzer—makes a sound.
- Motor—turns electricity into a spinning motion.

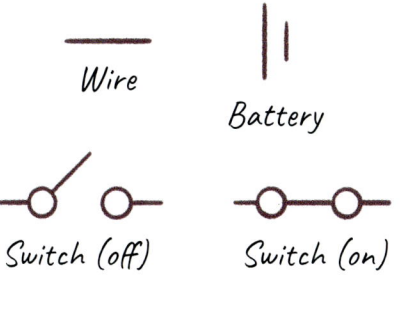

Wire | Battery | Switch (off) | Switch (on) | Buzzer | Motor | Bulb

SOUND PAGE 32

Parallel and series

Circuits can work in two different ways.

Two bulbs in series

Two bulbs in parallel

In a series circuit, all the parts of the circuit are on the same single loop.

In a parallel circuit, the loop splits into two or more pathways, and the current flows along both or all of them.

CURRENT ELECTRICITY PAGE 41

Current, voltage, and resistance

Current, voltage, and resistance are three important features of any circuit.
- Current is the flow of electricity.
- Voltage is the difference in charge between the two terminals. It's the "push" that makes electricity flow.
- Resistance is anything that slows down the current. For example, a bulb provides resistance since it contains a narrower wire, and it's harder for the current to get through.

One way of understanding this is to imagine a circuit is a hose full of flowing water.

- Current is the flow of water.
- Voltage is how hard the water is being pushed along.
- Resistance is anything that slows down the flow—like someone standing on the hose.

ELECTRICITY PAGE 40

43

Generating electricity

Electricity is a form of energy, and energy cannot be made or destroyed. We can only turn one type, or form, of energy into another. So we can only get electricity from other forms of energy.

The main way we make electricity is using a generator. It's a device that turns kinetic energy, or movement, into a flow of electricity in a wire.

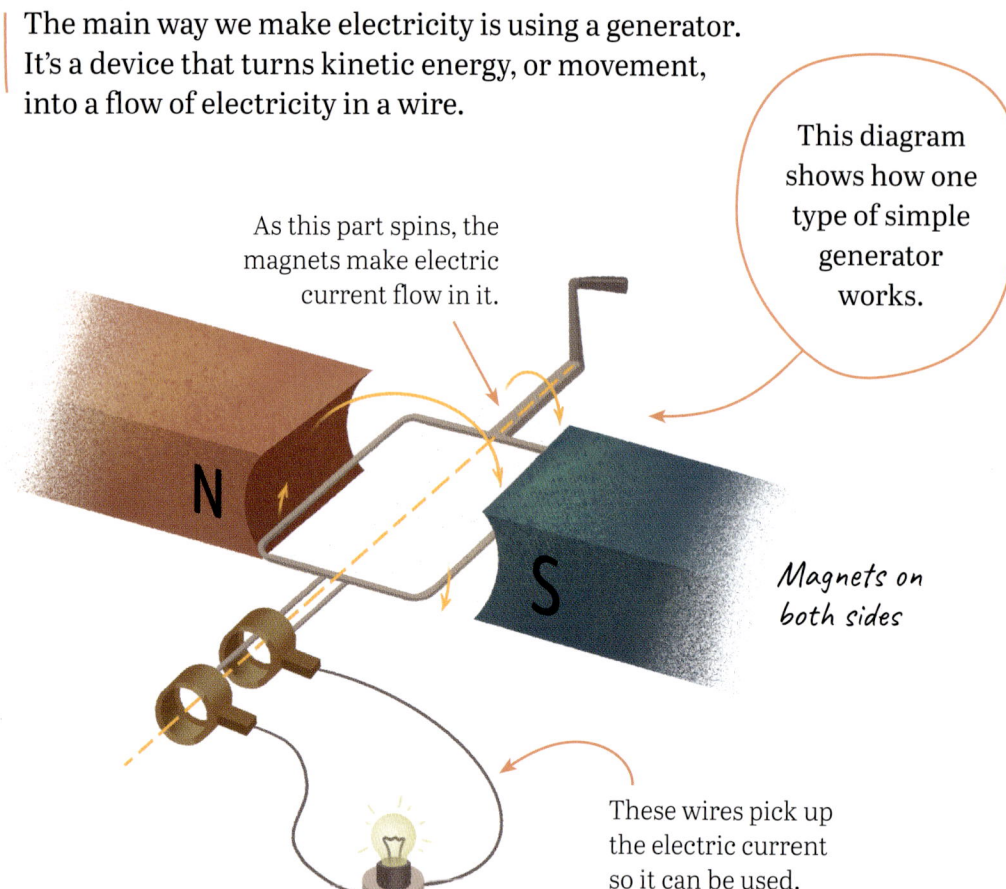

This diagram shows how one type of simple generator works.

As this part spins, the magnets make electric current flow in it.

Magnets on both sides

These wires pick up the electric current so it can be used.

Generators usually work using a spinning motion. For example, a wind turbine rotates in the wind, and has a generator inside to turn this motion into electricity. Most power plants work in a similar way.

Burning fuel

Traditional power plants burn fuel, such as oil, coal, or gas. The heat boils water, which turns into steam. The flow of steam works like wind, making a fan-shaped part called a turbine spin around. This drives the generator, making electricity flow.

However, burning fuel releases carbon dioxide as a waste gas, adding to global warming. So we are now switching to other ways of generating electricity that cause less pollution.

HEAT ENERGY
PAGE 22

44

Renewable energy

We're surrounded by moving wind, rivers, and seas, and we can use this natural movement to generate electricity. It's called "renewable" energy because, unlike fuels, these energy sources don't run out.

- **Wind power**—uses wind to make blades turn and power a generator.
- **Hydroelectric power**—uses the downhill flow of water, pulled by gravity, to make turbines spin.
- **Tidal power**—holds back the tide using a seawall, then lets the water flow through turbines to make them spin.

GRAVITY PAGE 52

Solar panels

Solar panels work in a different way. They're made of materials that can absorb (soak up) light energy, mainly from the Sun, and turn it into a flow of electricity.

- Solar panels on house roofs provide an electricity supply for the home.
- Some small gadgets are powered by a mini solar panel.
- There are also solar power plants made up of thousands of panels.

LIGHT PAGE 34

Batteries

Batteries contain chemicals that react together to turn chemical energy into electricity. We use them in many things, like gadgets, tools, toys, phones, and electric vehicles.

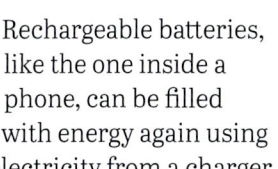

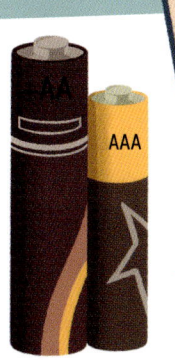

Disposable batteries stop working when the chemical energy is used up.

Rechargeable batteries, like the one inside a phone, can be filled with energy again using electricity from a charger.

CHEMICAL ENERGY PAGE 28

Electricity supplies

Most homes, especially in developed countries, have their own electricity supply. But how does it get there?

Bringing electricity into your home takes a huge network of power plants and generators, cables and wires. It's known as the grid, electric grid, or power grid.

The electric grid of a country is very complicated.

It usually has several different types of power plants and generators to generate the electricity.

They are linked to cables that deliver the electricity supply to the places where it's needed ...

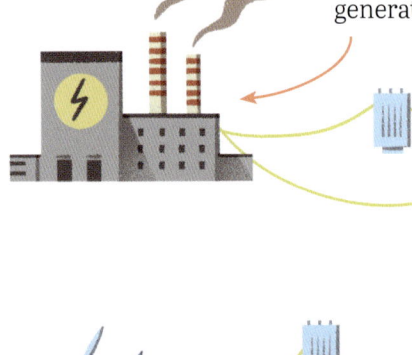

... not just to homes, but to schools, factories, hospitals, farms, streetlights, sports stadiums, railroads, and many more places.

Electricity can be dangerous to humans. If a powerful current flows through the body, it can cause serious burns or even be deadly. So we have to make sure that the systems for carrying electricity around are safe.

Delivering electricity

To carry the electricity long distances, we use high-voltage cables that carry a lot of electrical energy at once. For safety, they are carried on very high pylons, or buried underground.

Pylon

Cables

Insulating materials such as porcelain or glass stop the electric cables from touching the pylons, so the electricity can't flow down to the ground.

🔍 CURRENT ELECTRICTY PAGE 41

46

Electricity substations

The voltage in the cables is too high to use in your home. So first, the electricity flows into a substation, where the voltage is reduced. From there, it flows into underground electricity cables that run along the streets.

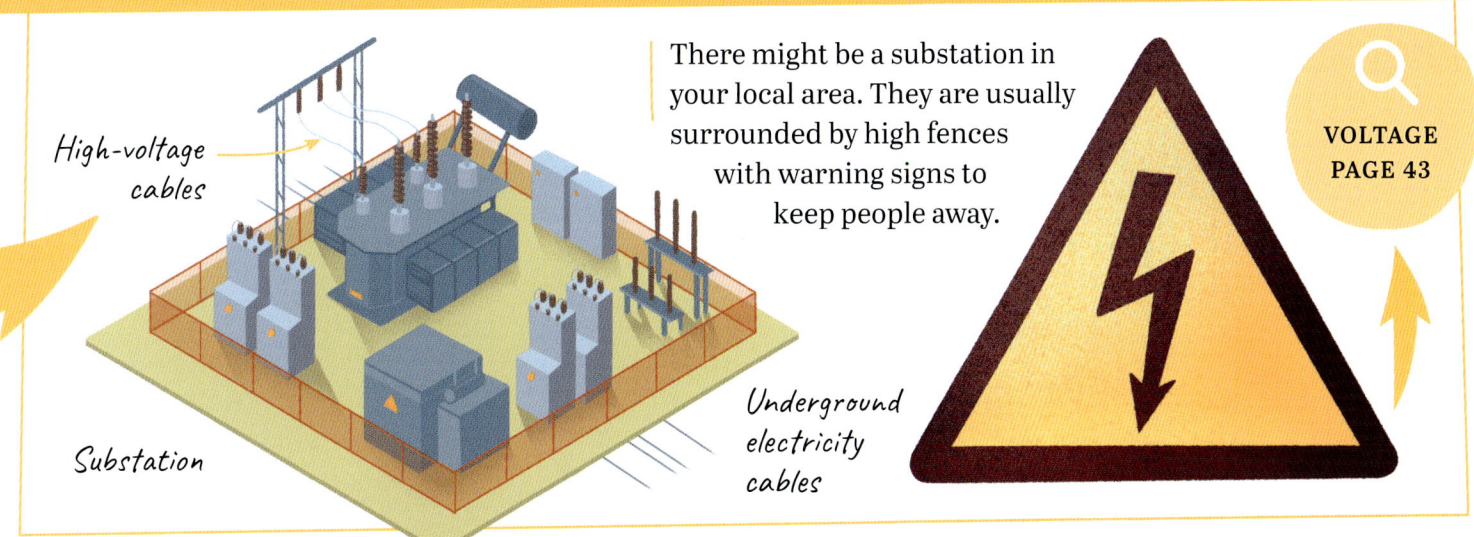

There might be a substation in your local area. They are usually surrounded by high fences with warning signs to keep people away.

VOLTAGE PAGE 43

In your home

From the power grid supply, cables branch off to connect to houses and other buildings. Inside the house, electric circuits inside the walls lead to all the plug sockets and light switches.

SWITCHES PAGE 43

Staying safe

Inside a house, electrical wires, plugs, and sockets have covers made of an insulator such as plastic or rubber. The electricity can't flow through the insulator, so it stays safely in the wires or the other metal parts inside, and can't harm you.

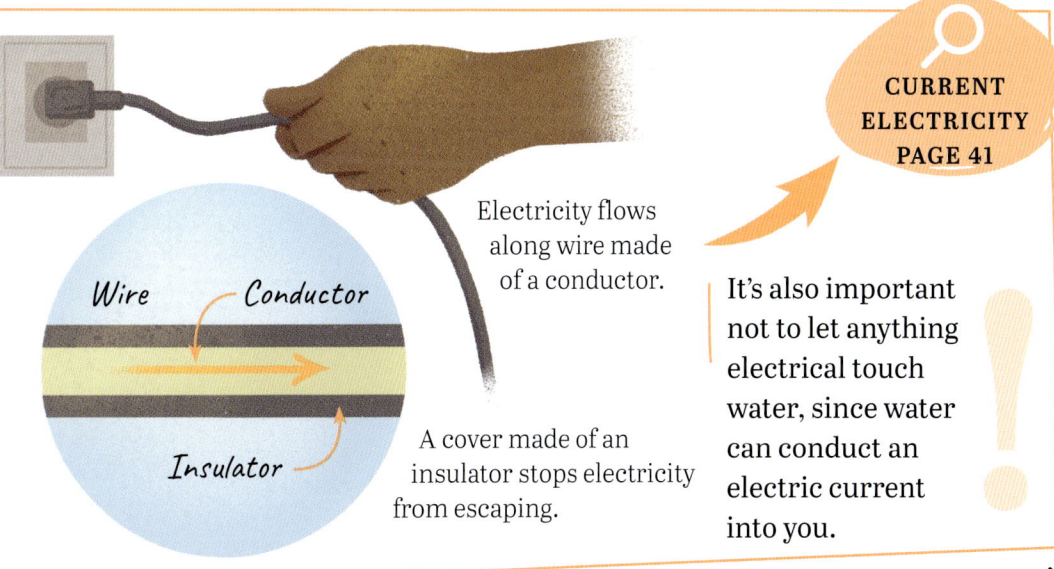

Electricity flows along wire made of a conductor.

A cover made of an insulator stops electricity from escaping.

CURRENT ELECTRICITY PAGE 41

It's also important not to let anything electrical touch water, since water can conduct an electric current into you.

47

Electronics

Electronics means using the flow of electricity in a circuit to do useful things. Instead of just providing power, an electronic circuit can carry signals and store information.

For example, computers work using electronic circuits, and so do many other modern inventions.

A computer like this laptop contains very small, complex electronic circuits. They use the flow of electricity to store and change information, and to display information on the screen.

In most modern electronic devices, the circuits are very small or even microscopic. This allows them to fit into a small object, like a laptop or smartphone, but still be able to hold a lot of information.

Tiny chips

Today, most electronic devices contain chips. A chip is a tiny slice of a material called silicon, covered in microscopic electronic circuits.

Instead of connecting lots of tiny wires together, the circuits are printed onto the chip. They are made of materials called semiconductors. Different arrangements of semiconductors make tiny switches and other parts that control the way electricity flows around the circuit.

MATTER
PAGE 68

Zeros and ones

In an electronic circuit, information is stored in binary code, a code made up of just two numbers, **0** and **1**. The tiny switches in a circuit can be switched off or on, and these two states stand for **0** and **1**.

ELECTTRIC CIRCUITS PAGE 42

For example, **10101101** is binary code for the number **173**

Information stored in circuits in this way is often called "digital" information.

Digital information

Although it sounds amazing, almost every kind of information can be stored in binary code as digital information, if you use enough **0**s and **1**s.

- Numbers

- Computer programs, which give computers instructions

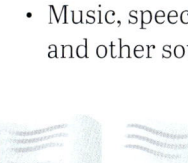
- Music, speech, and other sounds

- Videos

- Pictures and photos

- Letters, words, and books

SOUND PAGE 32

Digital world

The first electronic circuits were developed in the early 1900s. Since then, the rise of computing and digital storage has changed the world. For example, we can now send instant messages all over the world using email, texts, and messaging apps.

Electronics has changed these things, too …
- We can now find all kinds of information online, using the Internet.
- Money is stored as digital information, so you can do banking and go shopping online.
- Music, photos, and videos can all be stored as digital information, and played back on phones and computers.
- Electronic circuits are used in inventions like robots, smart speakers, and electronic musical instruments.

Long ago, people had to carry handwritten letters long distances.

Thanks to computers and smartphones, we can message anyone in an instant.

CIRCUITS PAGE 42

Pushes and pulls

Forces are a very important part of physics. They are pushes and pulls that affect objects and materials, making them move, stop moving, or change in some way.

There are several different types of forces, and there's usually more than one force acting on an object at any time. Sometimes, different forces balance each other out.

For example, when you throw a ball, your hand applies a pushing force that makes the ball move, and fly through the air.

Movement

Pushing force

More examples of forces:

After you throw a ball, gravity pulls it down to the ground.

Squeezing modeling clay makes it change shape.

A bungee cord stretches, then springs back with a pulling force.

Bicycle brakes slow down a bike's wheels and make the bike stop.

Contact forces

Contact forces happen when objects or materials touch each other—like when you throw a ball, pull a door closed, or press the keys on a piano.

Contact force

Friction is a contact force, too. It happens when objects or materials scrape, rub, or grip together.

FRICTION PAGE 54

Friction between walking shoes and rock keeps you from sliding downhill.

Forces at a distance

Some forces can work "at a distance." Imagine dropping a pebble—it will get pulled toward the ground by gravity, even though the ground is not touching it.

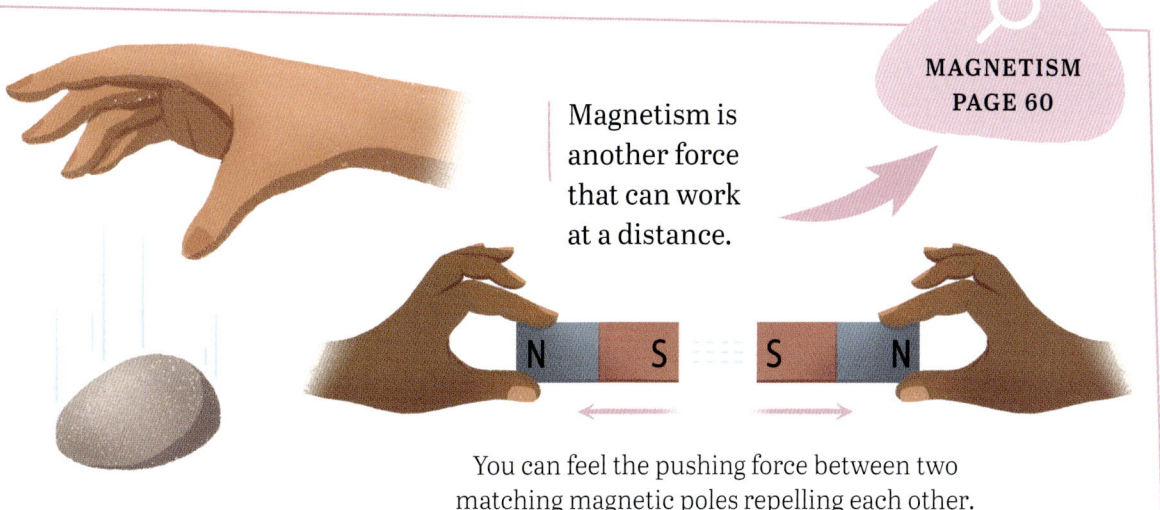

Magnetism is another force that can work at a distance.

MAGNETISM PAGE 60

You can feel the pushing force between two matching magnetic poles repelling each other.

Lots of forces

There are often several different forces acting on an object. What happens to the object depends on how the different forces affect it and how strong each force is.

RESISTANCE PAGE 55

For example, when a plane is flying through the air, these forces are acting on it.

Lift is a pushing force that pushes the plane up, thanks to air being pushed down by its wings.

The plane has enough thrust and lift to overcome the forces of drag and gravity, so it keeps flying forward.

Drag is friction with the air, which slows the plane down.

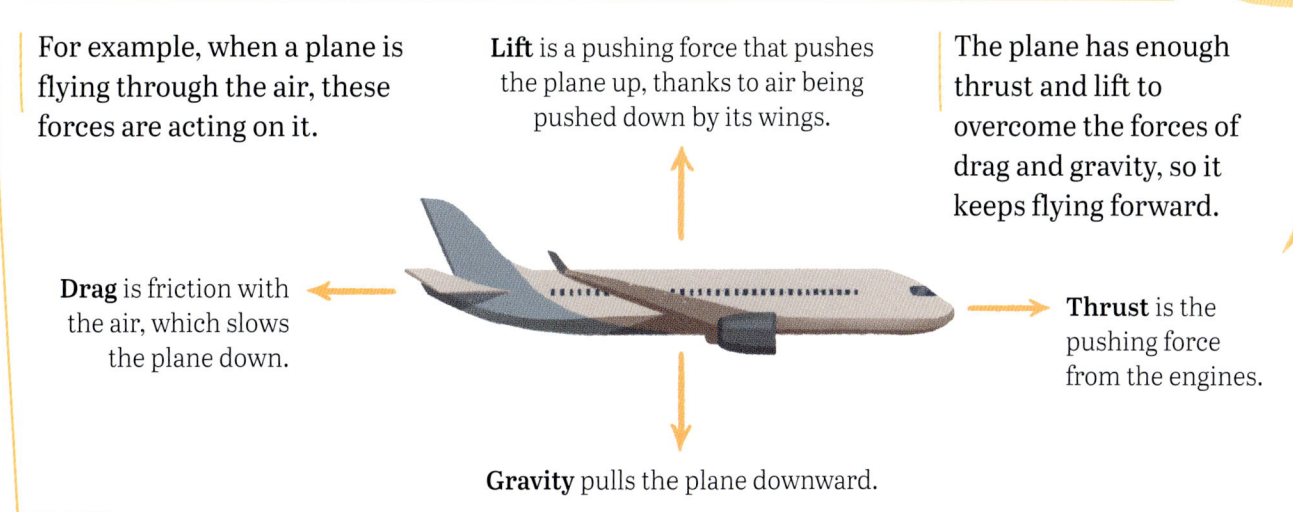

Thrust is the pushing force from the engines.

Gravity pulls the plane downward.

Balanced and unbalanced forces

Sometimes, two forces balance each other out.

A book on a table is being pulled down by gravity—but it's also being pushed up by the table.

These two tug-of-war teams are both pulling as hard as each other.

Since the two pulling forces are balanced, the rope doesn't move.

The forces are balanced, so the book stays still.

GRAVITY PAGE 52

Gravity

Gravity is a pulling force between objects. It exists when objects are in contact with each other, but it can also work at a distance.

When you think about gravity, you probably think about the way the Earth pulls things down. In fact, though, all objects have their own gravity.

We are mainly affected by the pull of the Earth's gravity because the Earth is a very large object with a lot of mass. The more mass an object has, the stronger its gravity is.

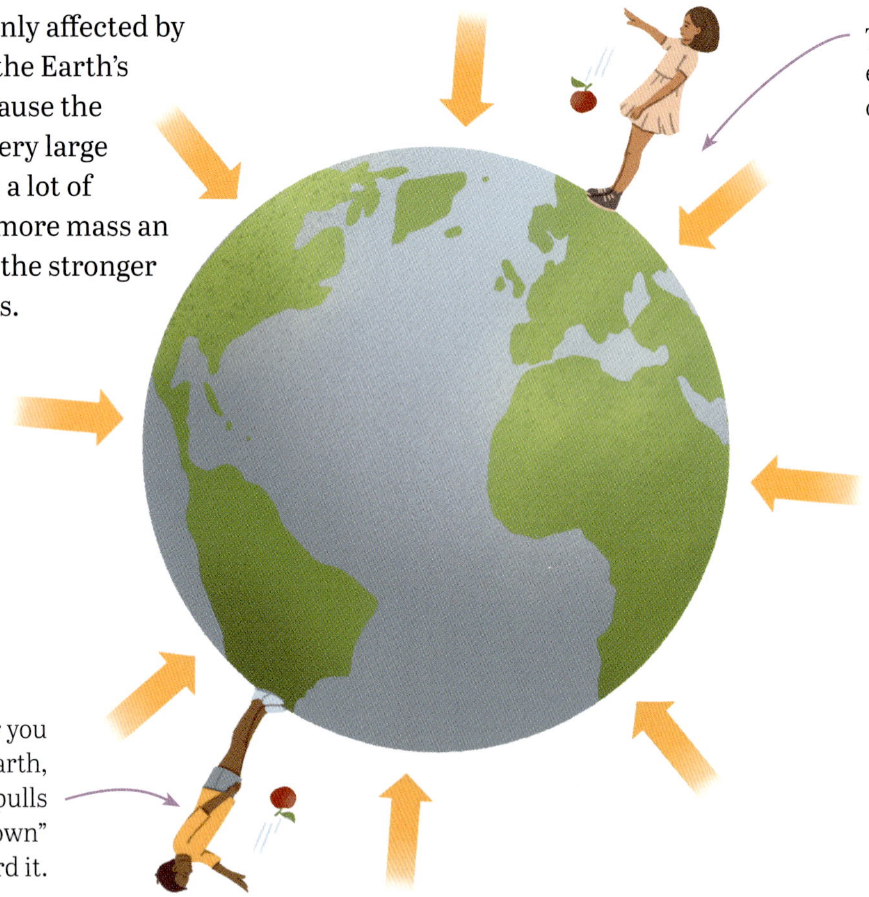

The Earth's gravity pulls everything toward its central point.

Wherever you are on Earth, its gravity pulls things "down" toward it.

Gravity can be hard to understand, and even scientists don't really know exactly how it works. But here's what we do know...

Gravity in other objects

Though other objects also have gravity, smaller objects like people and buildings have such weak gravity that we don't notice it. But we are affected by the Moon's gravity, which pulls on the Earth's seas and oceans, causing tides.

THE MOON
PAGE 84

Force fields

How can gravity work at a distance, and even across huge distances in outer space? It's because each object has an area of pulling force around it, called its gravitational field.

Earth's gravitational field stretches out into space all around it. But the farther away from Earth you are, the weaker the gravity is.

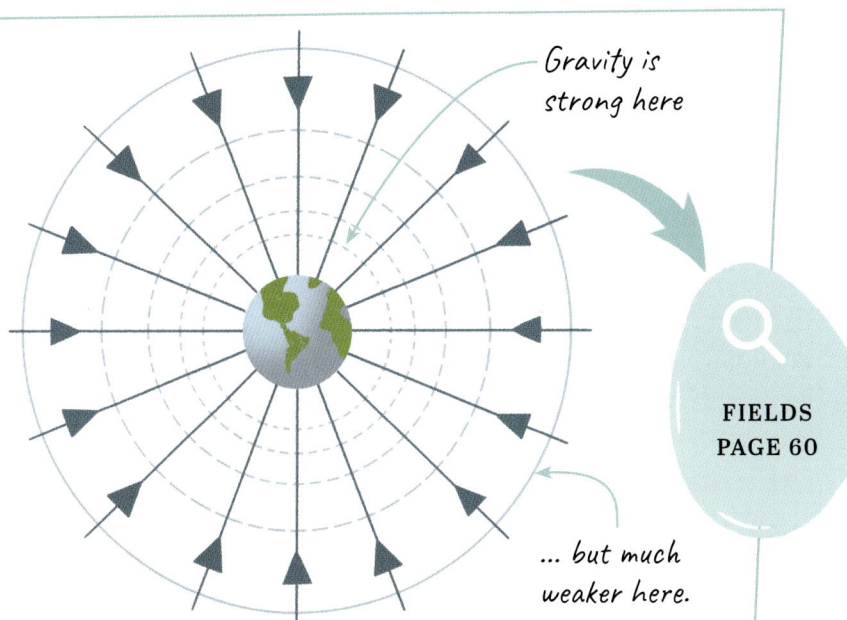

Gravity is strong here

... but much weaker here.

FIELDS PAGE 60

Gravity, mass and weight

If you pick up a big bag of groceries, it feels heavy, and you can feel its weight pulling down on you. But weight only exists because of gravity.

Mass is the amount of matter in an object, and it is measured in kilograms.

This bag of shopping has 10 kg (22 lb) of mass.

On Earth, the bag weighs 10 kg (22 lb). But if it was on the Moon, it would weigh less: about 1.66 kg (3.7 lb), since the Moon has much weaker gravity.

And if it was in outer space, far from any stars or planets, it would float around with hardly any weight.

SPACE PAGE 80

Falling objects

On Earth, gravity is always pulling on objects. If no other force stops them, objects will fall to the ground.

A falling object starts slowly.

As an object falls, it accelerates, or speeds up, and falls faster and faster.

ACCELERATION PAGE 65

But as it falls, gravity keeps pulling on it, making it fall faster.

Time	Falling speed
0 seconds	0 meters per second
1 second	about 10 meters per second
2 seconds	about 20 meters per second
3 seconds	about 30 meters per second

That's why falling a small distance is usually safe, but falling a long way is more dangerous.

Friction

Friction is a dragging, scraping, or rubbing force. It slows down or stops objects or materials when they slide or push past each other.

We experience friction all the time in everyday life. It makes brakes work, makes things wear out, and makes different surfaces grip together.

Everyday examples of friction:

It's hard to drag heavy furniture along the floor, because friction slows it down.

Textured silicone oven gloves use friction to grip dishes.

Friction is an important part of physics, because it affects how almost all the objects and materials around us move and behave. It can be useful, but it can also cause problems.

How friction works

Friction happens when materials rub, scrape, or flow past each other.

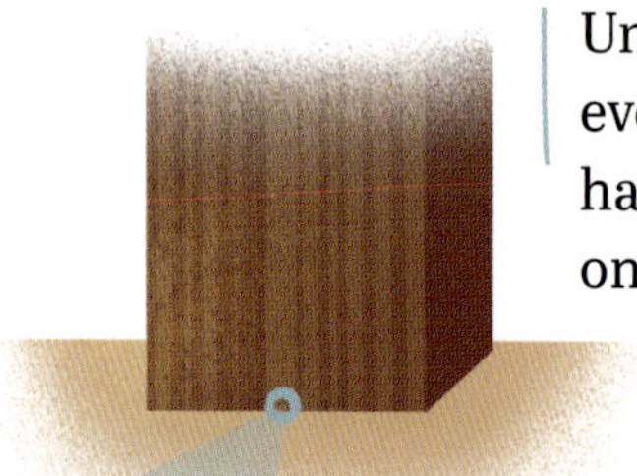

Under a microscope, even smooth materials have bumps and lumps on the surface.

Some surfaces have much more friction than others. Sometimes, friction completely stops movement, like when brakes keep a wheel from turning.

But if there is less friction, surfaces can slide past each other, like ice skates on ice.

When two surfaces rub together, the tiny bumps scrape and catch onto each other.

MATTER
PAGE 68

54

Resistance

Resistance, also called drag, happens when an object is moving though a liquid or gas, such as water or air.

For example, as a submarine moves through water, friction with the water flowing past it slows it down.

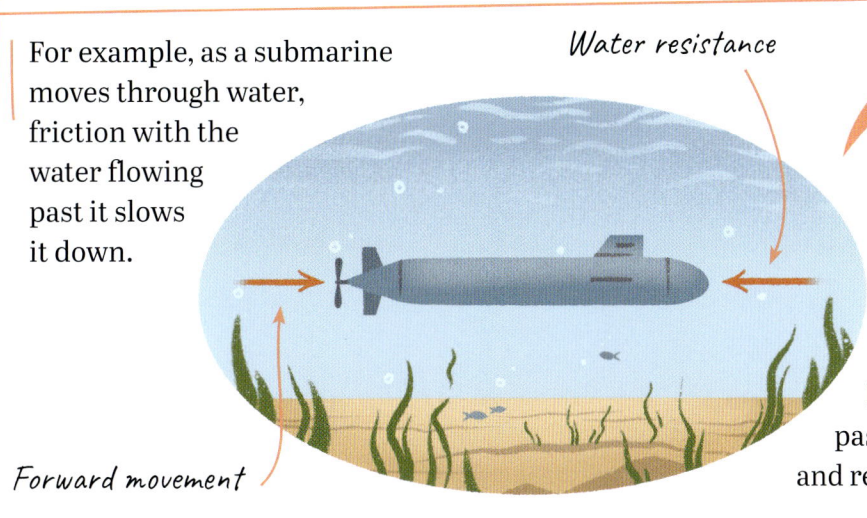

Water resistance

Forward movement

LIQUIDS AND GASES PAGE 70

Aircraft, boats, submarines, and some animals have a narrow, pointed shape called a streamlined shape. This helps air or water to flow past them more smoothly, and reduces resistance.

Good and bad friction

We actually need friction to make most of the things around us work. Without friction, life would be very difficult!

- You wouldn't be able to grip everyday objects like a water bottle or pencil.
- You couldn't even stand still. You would slide around as if you were on very slippery ice.
- Cars, trains, and bikes wouldn't work, since they wouldn't be able to grip the ground.

ICE PAGE 71

However, friction also causes problems, like making things wear out and get stuck.

- It wears holes in your jeans.
- It makes engine parts wear out.
- It wears down car tires.

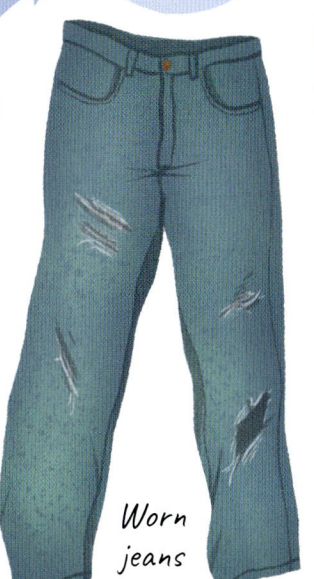

Worn jeans

Friction and heat

Friction has another important effect—it heats things up. When two materials rub together, it makes their atoms and molecules move faster, and they get warmer.

That's why rubbing your hands together when you're cold warms them up.

It's also why you can start a fire by rubbing sticks or pieces of wood together—but they have to be moving very fast for it to work.

HEAT ENERGY PAGE 22

Springs and elastic

Stringy and elastic objects and materials can store energy and then release it, creating a pushing or pulling force.

Springs and elastic are very useful, and you can find them in many everyday objects.

A diving board is a kind of spring. When you jump on it, it bends, storing energy. Then it springs back up with a pushing force, helping you jump higher into the air.

Pushing force bends board down

Board springs back and pushes up

BOINGGG!

When you stretch a rubber band, it pulls back with a pulling force—so you can use it to hold things together tightly.

Rubber band

How springy or elastic something is depends on the material it's made of, and its shape. Different kinds of elastic and springy things work in different ways.

How it works

All solid materials contain atoms and molecules that have forces holding them together. When you bend or stretch something, you pull them apart slightly.

🔍 ATOMS AND MOLECULES PAGE 68–69

In many materials, this will quickly make them break or change shape. But some materials can bend or stretch more, and then spring back to their original shape. They include rubber, some types of plastic, and many metals.

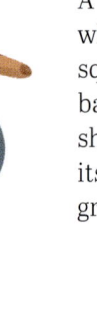

A rubber ball bounces because when it hits the ground, it gets squeezed. Then it pops back to its round shape, pushing itself up off the ground again.

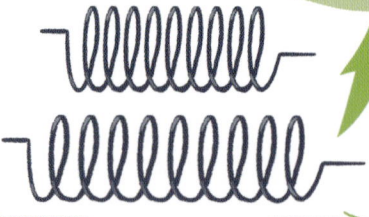

A metal spring is a coil of metal in a spiral shape. When it gets pulled, each part of the spring bends only a tiny bit, but the whole spring can stretch quite a lot, making it very elastic and springy.

56

Deformation

When a spring, bouncy ball or other elastic object changes shape, it's called deformation. There are several different ways this can happen.

- **Squeezing** or **compression**—like when a bouncy ball gets squashed as it hits the ground.

- **Stretching** or **tension**—like when you pull on a spring, making it longer.

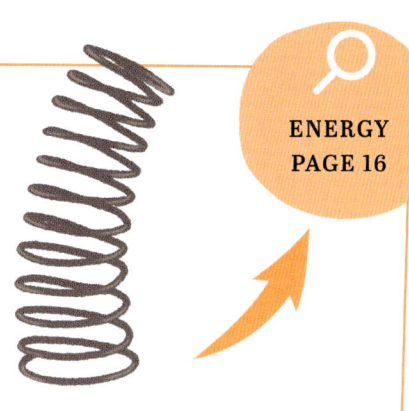

ENERGY PAGE 16

- **Bending**—like when you jump on a diving board.

- **Twisting**—some elastic objects, like a rubber band, can be twisted to store energy.

Springy gases

Usually, elastic things are solid. However, air and other gases can also act like a spring.

For example, think about a bouncy castle. It's pumped full of air to make it springy.

The molecules in the air get pushed closer together.

When you jump on it, you push down and squeeze the air inside.

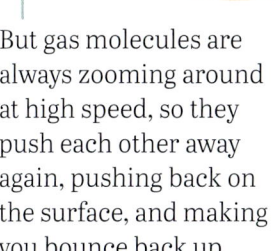

But gas molecules are always zooming around at high speed, so they push each other away again, pushing back on the surface, and making you bounce back up.

GASES PAGE 70

The elastic limit

Even very stretchy and flexible materials and objects can't keep stretching or bending forever. They have a limit to how much they will deform, which is called the elastic limit.

If you pull a spring beyond its elastic limit, it won't bounce back. Instead, it will stay stretched out and stop working.

MATTER PAGE 68

SNAP!

And if you bend a flexible ruler too far, it will break.

Pressure

Pressure means how much force is acting on a particular area of a surface or object.

The smaller the area a force is pushing on, the greater the pressure will be on that area.

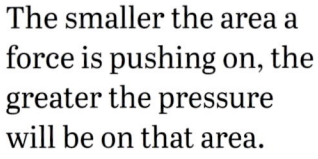
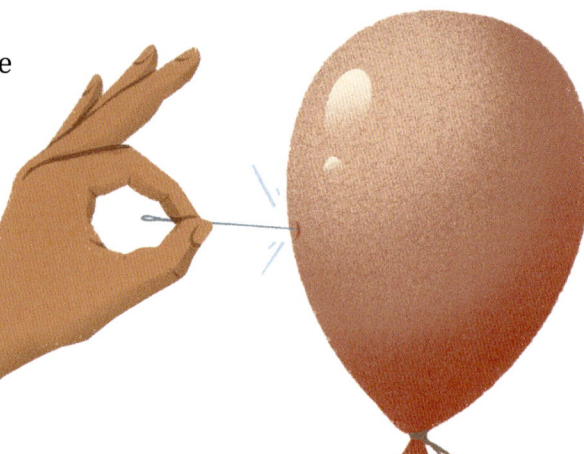

The very small area of the sharp pinpoint concentrates the force into much more pressure, enough to break the balloon skin.

For example, if you press a balloon with your hand, it will squash it slightly. But if you use the exact same amount of force, but hold a pin in your hand and press the balloon with that instead, the balloon will pop.

The physics of pressure helps to explain how many everyday items and situations work, from pins, vegetable knives, and scissors to how boats float.

Pressure numbers

To calculate pressure, physicists use a simple equation.

$$\text{Pressure (P)} = \frac{\text{Force (F)}}{\text{Area (A)}}$$

In other words, the pressure is the force divided by the area.

Here's an example. Imagine a person weighing 50 kg (110 lb) is standing on stilts. The stilts cover an area of 5 sq cm (¾ sq in) on the ground.

50 kg of force

Pressure = 50 divided by 5, or 10 kg per square cm

5 square cm (¾ square in) of area

FORCES PAGE 50

Atmospheric pressure

Everyone on Earth is being squeezed all the time by pressure from the atmosphere—the layer of gases that surrounds the Earth. We don't notice it, because we're used to it and our bodies are built to cope with it.

The pressure is caused by the weight of all the air pressing down on us. Because it's made of gas and is not solid, the air flows all around us and presses on us from every direction.

GASES PAGE 70

Higher up, such as on the top of a high mountain, the atmospheric pressure is lower.

Closer to Earth, the particles of air are squashed closer together by the weight of air on top of them, and the pressure is greater.

Water pressure

Water and other liquids also put pressure on objects. If you swim at the surface of the sea, you won't feel much extra pressure. But the deeper down you go, the greater the pressure is.

A sphere is the best shape for the cabin, since it's equally strong in all directions.

LIQUIDS PAGE 70

Deep-sea submersibles, which explore the deep oceans, have to be very strong to resist the water pressure and keep the crew safe.

Deep-sea fish and other creatures have adapted to cope with the pressure, so it doesn't harm them.

Floating

Atmospheric pressure and water pressure push in all directions, including upward. When an object floats in water, it is being pushed up by the water under it.

If an object is denser (heavier for its volume) than water, the upthrust can't support it, so it sinks. But if it's less dense than water, the upthrust is stronger than the downward force of gravity pulling on the object's weight, so it floats.

GRAVITY PAGE 52

The force of the water pressure pushing up is called upthrust.

59

Magnetic and electric forces

Along with gravity, magnetic and electric forces are "at a distance" forces that can work across an empty space.

Like gravity, these forces work using fields. A field is an area or zone that reaches out around a magnet or something with an electric charge. The force works on objects that are within the field.

Magnetic force can pull some types of metal, known as magnetic metals, toward a magnet.

Separately, magnetic and electric forces can have different effects. But they can also work together as electromagnetism.

Electric force is what makes electricity flow around a circuit. It is created when an object has a positive or negative electric charge.

A charged battery creates an electric force, which makes an electric current flow around a circuit.

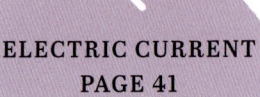

ELECTRIC CURRENT
PAGE 41

How magnets work

Magnets work because of the way their atoms are arranged.

METALS
PAGE 6

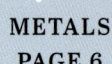

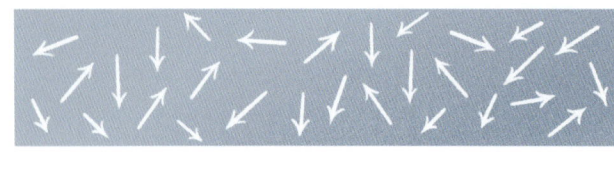

All atoms have tiny pulling forces that hold them together. But in most materials, the forces all point in different directions, so they cancel each other out.

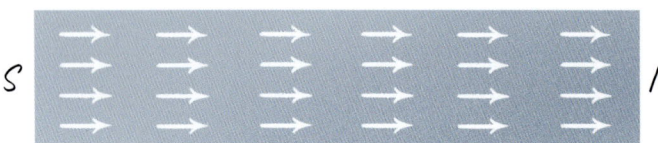

In a magnet, the forces are lined up in the same direction, so they create a larger pulling force.

Magnets pull on magnetic metals, such as iron, nickel, and cobalt. These metals contain atoms that can all be pulled in the same direction by the magnet.

60

Magnetic forces

A magnet has two opposite ends or sides, called poles. One is the north pole, and one is the south pole.

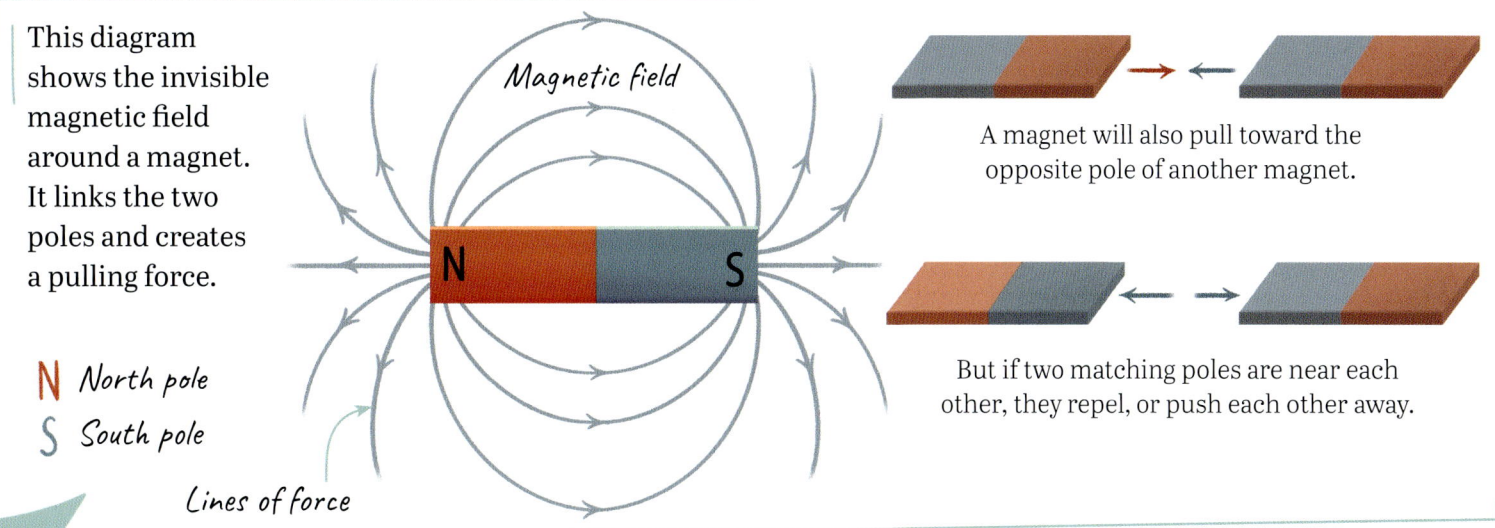

This diagram shows the invisible magnetic field around a magnet. It links the two poles and creates a pulling force.

N North pole
S South pole

Lines of force

A magnet will also pull toward the opposite pole of another magnet.

But if two matching poles are near each other, they repel, or push each other away.

Electric force

An electric field is very similar to a magnetic field.

+ Positive charge
− Negative charge

Electric field

Lines of force

ELECTRIC CHARGES PAGE 40

An electric field forms between an object with a positive charge and one with a negative charge. It makes charge flow from one to the other.

Electromagnetism

Electricity and magnetism are closely related. A changing electric field creates a magnetic field, and a changing magnetic field creates an electric field. Because of this, they are often described as one thing—electromagnetism.

An electric field can be used to turn a piece of metal into a magnet that can be switched on and off, called an electromagnet.

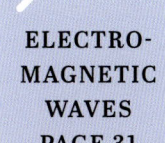

ELECTRO-MAGNETIC WAVES PAGE 31

All of space is full of electric and magnetic fields. Electromagnetic waves such as light travel as vibrations in these fields.

Laws of motion

In the 1600s, the great scientist Isaac Newton discovered three important laws, or rules, about how motion works.

Isaac Newton was an English scientist who studied physics, chemistry, space, and many other things.

He made important findings and theories about motion, gravity, and light. He wrote a famous book, known as the *Principia*, that sets out many of the things he discovered.

Isaac Newton

Newton's three laws are simple, but they were new ideas back then. They changed the way people understood physics.

Newton's first law

Newton's first law of motion says that an object will keep doing what it's doing, unless a force acts on it to change that. The way an object keeps doing the same thing is called inertia.

For example, a still ball will stay still, unless something pushes or moves it.

And a moving object will keep going in the same direction and at the same speed, unless a force acts on it to change it.

When you kick a ball, it will keep going forever in the same direction, unless other forces act on it.

On Earth, a ball will curve through the air, fall to the ground, and stop rolling, because other forces are acting on it—air resistance, gravity, and friction with the ground.

FRICTION PAGE 54

Newton's second law

Newton's second law of motion says that the more force there is acting on an object, the faster it will accelerate, or speed up.

This means that people fall more slowly on the Moon than on Earth, since the force of gravity is weaker.

However, the more mass an object has, the more slowly it will accelerate, given the same amount of force.

GRAVITY
PAGE 52

That's why it's easier to push a toy car, and harder to push a full-sized car.

Newton's third law

Newton's third law of motion says that for every action, there is an equal and opposite reaction. This just means that when a force acts on an object, the object pushes back with the same force.

For example, if someone on roller skates pushes another person, they will both move apart.

PUSHES AND PULLS
PAGE 50

Dynamics

The physics of how forces affect motion is called dynamics. It's very important for things like designing machines, planning space missions, and making sure things like parachutes and roller-coasters are safe.

The science of dynamics is used to calculate how big a parachute needs to be and how fast it will fall.

ROLLER-COASTERS
PAGE 15

63

Measuring motion

Physicists have several ways to describe and measure motion, using different methods and units of measurement.

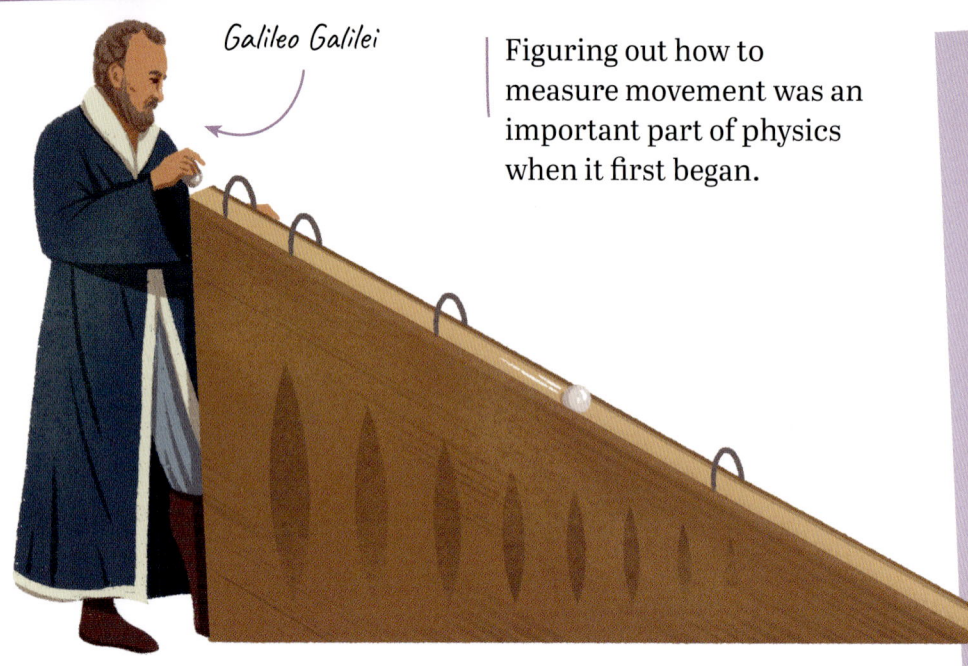

Galileo Galilei

Figuring out how to measure movement was an important part of physics when it first began.

Italian scientist Galileo Galilei is famous for his motion-measuring experiments in the 1600s. In one, he measured acceleration by letting a ball roll down a slope, and timing how long it took to reach different points.

Physicists don't always use the same units of measurement that we do in everyday life. For example, most people would measure speed in kilometers per hour or miles per hour, but physicists often use m/s (meters per second) or ft/s (feet per second) instead.

Speed

Speed means how fast something is moving. It's measured as a distance per unit of time—for example, miles per hour or meters per second.

An object's speed is the same whether it's going in a straight line, changing direction, or moving around in a circle.

🔍 TIME PAGE 11

If a bird is flying at 10 m/s (33 ft/s) that means in one second it moves 10 m (33 ft), in 5 seconds it moves 50 m (164 ft), and so on.

This toy train is moving at 0.2 m/s (0.7 ft/s) as it goes around and around the track.

Velocity

Velocity is slightly different from speed. It's a measurement of how fast something is moving in a particular direction.

This car is driving eastward at a speed of 30 km/h (19 mph).

Its velocity in an eastward direction is 30 km/h (19 mph).

When it turns north, its speed stays the same.

But its velocity is now zero in an eastward direction, since it is not moving east at all.

🔍 NORTH AND SOUTH PAGE 79

Acceleration

Acceleration means speeding up. It's harder to measure and describe, because the speed of an accelerating object is always changing.

The unit of measurement for acceleration is m/s^2, or meters per second (ft/s^2 or feet per second).

🔍 ACCELERATION PAGE 53

This plane is accelerating.

Here, it's going at 100 m/s (328 ft/s)

One second later, it's going at 110 m/s (360 ft/s)

2 seconds later 120 m/s (394 ft/s)

3 seconds later 130 m/s (426 ft/s)

4 seconds later 140 m/s (460 ft/s)

Every second, the plane goes 10 m/s (33 ft/s) faster than it was before. This is an acceleration of 10 m/s (33 ft/s) per second, or 10 m/s^2 (33 ft/s^2).

Momentum

Moving objects have something called momentum, which is their velocity multiplied by their mass. It's measured in kg·m/s (lb·ft/s).

The more momentum an object has, the more force it takes to make it stop.

🔍 MASS PAGE 53

1 kg·m/s is the momentum of 1kg of mass moving at 1 m/s. This 1 kg ball has a velocity of 10 m/s. So its momentum is 10 kg·m/s (72.3 lb·ft/s).

45 m

81 m

That's why it takes bigger, heavier vehicles longer to slow down and stop when they brake.

65

Simple machines

Machines are inventions that make it easier to do tasks by changing the way forces work.

Machines don't have to be big and complicated, or have lots of moving parts. Simple objects can be machines.

It would be hard to get this heavy crate into the back of this van.

You might need two people to lift it.

Simple machines are found in all kinds of everyday tools and gadgets, as well as making up some of the parts in more complicated machines.

But if you use a ramp, it's easier.

The crate moves farther, so the force you need is spread out over a longer distance, and it doesn't need as much strength.

A ramp, or "inclined plane," is an example of a very simple machine.

Wedges

A wedge is a shape that concentrates pressure into a very narrow area.

The sharp blade of a knife, an axe, or even a pair of scissors acts as a wedge.

The sharp edges turn a small amount of pressure from your fingers into a lot of force, so they can cut through paper and fabric easily.

🔍 PRESSURE PAGE 58

Levers

Levers are simple machines that can make forces stronger or weaker.

A lever is a rigid (non-flexible) object that rests on a fixed point called a fulcrum.

This basic lever can help to lift a heavy load.

When you push the long end down, the short end moves up.

The long end moves a longer distance, but you don't need to use much force.

Fulcrum

The short end moves a shorter distance, but with more force, so it can lift the heavy weight.

Pliers are made from two levers fixed together.

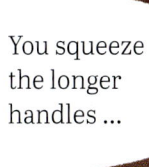

Fulcrum

You squeeze the longer handles ...

... and the shorter ends squeeze together very hard.

🔍 **WEIGHT PAGE 53**

Wheels and axles

Wheels are incredibly useful machines that were invented around 5,500 years ago.

🔍 **FRICTION PAGE 54**

A wheel is a flat, round shape that fits onto an axle, so that it can spin around.

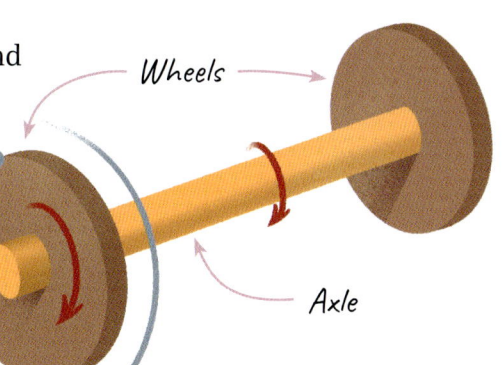

Wheels

Axle

By attaching a platform to an axle, you can carry a heavy load by rolling it along on wheels, reducing friction.

Axle

Gears

Wheels are also used inside bigger machines to link different parts together and pass on forces from one part to another.

Gears are wheels with "teeth" around the edge that can fit together. As one gear turns, it can turn another gear that is locked into it.

When a larger gear turns a smaller gear, the smaller gear spins faster.

A gear also makes the gear next to it turn in the opposite direction.

🔍 **FORCES PAGE 50**

67

What is matter?

Matter is the stuff all around us—wood, metal, plastic, water, air, and all the other substances and materials that everything is made from. But what is matter itself made of?

Matter is made of tiny, tiny units called atoms. On its own, each atom is much too small to see.

Atoms are spherical, like microscopic balls. They are so small that a teaspoonful of water contains 500 billion trillion atoms!

Different types of atoms, in different arrangements and mixtures, make up all the millions of different materials in the Universe.

Atom

And an average paperclip is made of around 1 billion trillion (or 1,000,000,000,000,000,000,000) atoms.

Atoms themselves are made of even smaller parts, or particles, as you can see here.

In the middle is the nucleus.

Neutron

Proton

Electron

Smaller particles called electrons zoom around the nucleus at high speed.

The nucleus is made of two kinds of particles: protons and neutrons.

Types of atoms

There are 118 different types of atoms. They are all slightly different sizes, and have different numbers of particles.

For example, a helium atom is small, with just two protons, two neutrons, and two electrons.

An oxygen atom is bigger, and has more mass, as it contains more particles. It has eight protons, eight neutrons, and eight electrons.

Helium

Oxygen

In atoms with more than two electrons, the electrons are arranged in layers, or "shells."

ELECTRONS PAGE 40

Elements

An element is a pure substance made of just one type of atom. Since there are 118 types of atoms, there are the same number of elements.

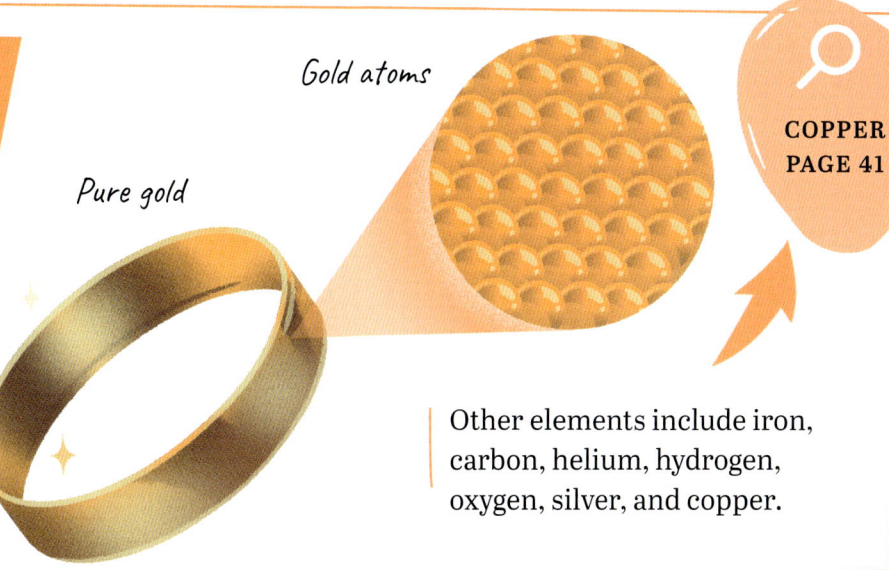

Gold atoms

Pure gold

COPPER PAGE 41

For example, gold is an element made of only gold atoms.

Other elements include iron, carbon, helium, hydrogen, oxygen, silver, and copper.

Making molecules

Atoms can join or "bond" with other atoms to make molecules. Lots of materials are "compounds" made of molecules rather than single atoms.

Water is a compound. Its molecules are made of two types of atoms: oxygen and hydrogen.

Some molecules are much bigger and more complicated.

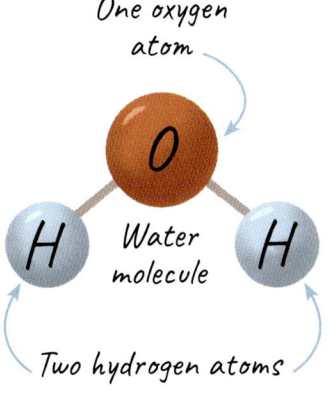

One oxygen atom

Water molecule

Two hydrogen atoms

This is a molecule of glucose, the sugar found in food.

Carbon atom

Oxygen atom

Hydrogen atom

FOOD PAGE 28

Making mixtures

Lots of everyday materials are mixtures. They are not pure elements like gold, or pure compounds like water, but contain a mixture of different things.

Brass is a mixture of the elements copper and zinc.

Seawater is one example. It's a mixture of water and salt, with small amounts of other substances.

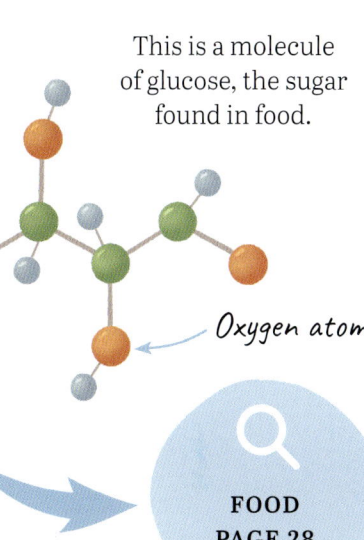

THE SEA PAGE 30

States of matter

Matter can exist in three main states: solid, liquid, and gas.

We're used to dealing with solids, liquids, and gases in everyday life, and the different ways they behave.

Solid

Solids keep their shape. If you put a brick in a container, it will stay the same shape.

Liquid

Liquids can flow, pour, and splash. In a container, a liquid such as water will spread out into an even layer with a flat surface.

Gas
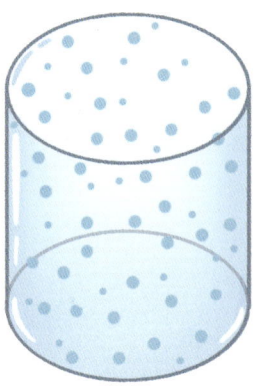
Gases are usually invisible, and spread out in all directions to fill the space they are in.

Solids, liquids, and gases behave the way they do because of the amount of energy in their atoms and molecules.

Moving particles

Atoms and molecules are always moving, but they can move at different speeds and have different amounts of kinetic energy, or movement energy.

In a solid, they are tightly packed together. They jiggle about but stay in a fixed pattern.

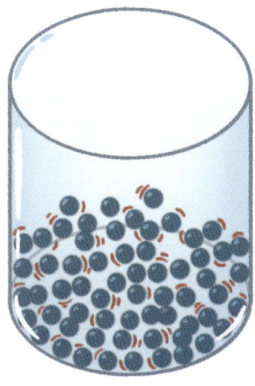
In a liquid, they are farther apart, move faster, and slip and slide past each other.

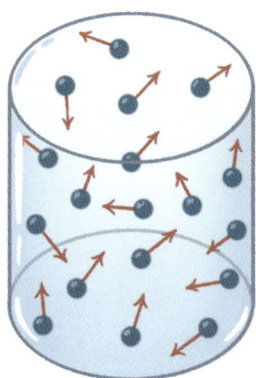
In a gas, they are even farther apart. They zoom around at high speed and bounce off each other.

🔍 KINETIC ENERGY PAGE 18

Changing state

A material's state of matter depends on its temperature. As the temperature changes, the state of matter can change, too. We often see this happening with water.

As ice warms up, it melts from a solid into liquid water, then evaporates into water vapor, a gas.

The opposite happens when it cools down. As water vapor cools, it condenses into liquid water, then freezes into solid ice.

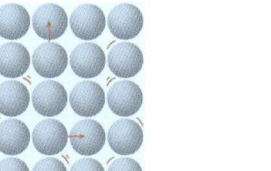
Solid ice

Melting → ← *Freezing*

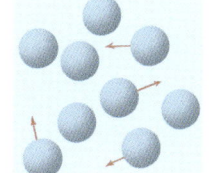
Liquid water

Evaporating → ← *Condensing*

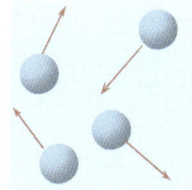
Water vapor (gas)

🔍 ICE PAGE 27

Different temperatures

We're used to water existing around us in all three states. For example, when we dry washing, water evaporates into a gas in the air. When we put an ice cube tray full of water in a freezer, it freezes solid.

But we don't see this with everything, because different materials change state at different temperatures. For example, rock is solid at normal, everyday temperatures. Most rocks only melt into liquid at very high temperatures of around 1,000 °C (1,832 °F).

Lava is melted rock.

We only see melted rock when a volcano erupts and hot liquid rock from inside the Earth bursts out.

🔍 INSIDE THE EARTH PAGE 78

Plasma

There is a fourth state of matter, called plasma. It can sometimes be created when gases get very hot, or electrically charged.

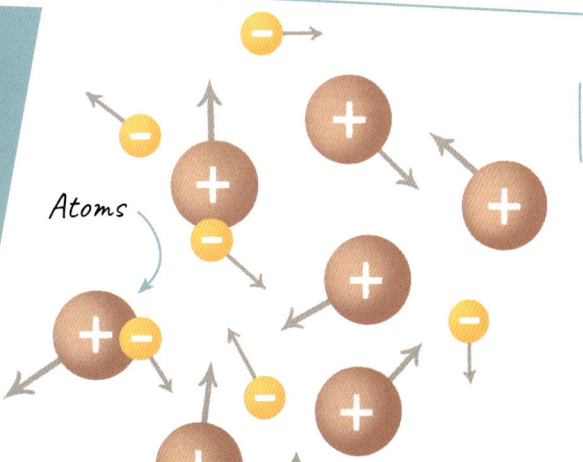
Atoms

In a plasma, the electrons break free from their atoms.

Free electrons

Plasma can be found in:
- Lightning strikes
- Stars, such as our Sun
- Neon lights

🔍 ELECTRIC CHARGES PAGE 40

71

Radiation and radioactivity

Radiation is energy that is given off by atoms. It includes electromagnetic energy, such as light and radio waves, but there are other types, too.

Radiation can take the form of waves or particles (tiny parts).

Visible light spectrum

🔍 ELECTROMAGNETIC SPECTRUM PAGE 38

Radio waves   *Microwaves*   *Infrared light*   *Ultraviolet light*   *X-rays*   *Gamma rays*

Radiation waves include all the waves in the Electromagnetic Spectrum, or EMS.

Radiation particles are tiny parts that fly out from an atom and can hit the things around them. They include alpha particles and beta particles.

An alpha particle is made of two protons and two neutrons.

Atom

A positron is similar to an electron but has a positive charge, while electrons have a negative charge.

A beta particle is a fast-moving, high energy electron or positron.

Some types of radiation are dangerous because they can damage other atoms, including the atoms that make up living things.

Ionizing radiation

Alpha particles, beta particles, gamma rays, and X-rays are forms of ionizing radiation. This is high-energy radiation that can knock electrons out of other atoms when it hits them.

Ionizing radiation

Electrons knocked out

🔍 X-RAYS PAGE 39

This creates ions, which are atoms with missing or extra electrons.

Damaging cells

Ionizing radiation can be harmful to living cells, because it breaks their atoms apart. The ions and free electrons can react with other atoms, forming new chemicals. This can kill cells or keep them from working as they should.

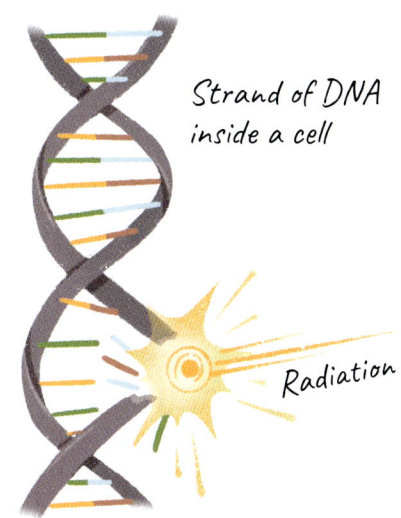

Strand of DNA inside a cell

Radiation

Sometimes, ionizing radiation damages DNA, the chemical inside cells that controls how they work. This can make living things grow differently from normal, or it can cause cancer when cells start growing out of control.

🔍 CHEMICALS PAGE 28

Radioactive atoms

Atoms that give out ionizing radiation are called "radioactive."

Most atoms have a balanced number of neutrons and protons. But sometimes, an atom can be unbalanced, or unstable. This makes it release energy in the form of ionizing radiation.

Elements with strongly radioactive atoms include:
- Uranium
- Plutonium
- Polonium
- Francium
- Radium
- Curium

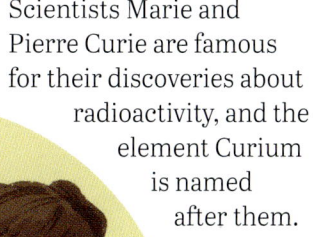

Scientists Marie and Pierre Curie are famous for their discoveries about radioactivity, and the element Curium is named after them.

🔍 URANIUM PAGE 75

Useful radiation

Though ionizing radiation can be harmful, it has many uses, too. For example:

BEEP BEEP BEEP!

- Gamma rays are used to kill mold and bacteria cells in food to make it last longer.
- Smoke alarms make a tiny amount of alpha particle radiation. Smoke blocks the particles, triggering the alarm.
- Beta particles are used in paper factories to test paper thickness. A machine measures how many beta particles can get through the paper, showing how thick it is.

🔍 GAMMA RAYS PAGE 39

Fission and fusion

Nuclear fission means splitting the nucleus (middle) of an atom apart. Nuclear fusion happens when two nuclei join together.

Both fission and fusion release energy. This is because of the size of the atoms involved.

In **nuclear fission**, a large atomic nucleus, usually a radioactive atom such as uranium, is split apart. After splitting, it needs less energy to hold itself together, so energy is released.

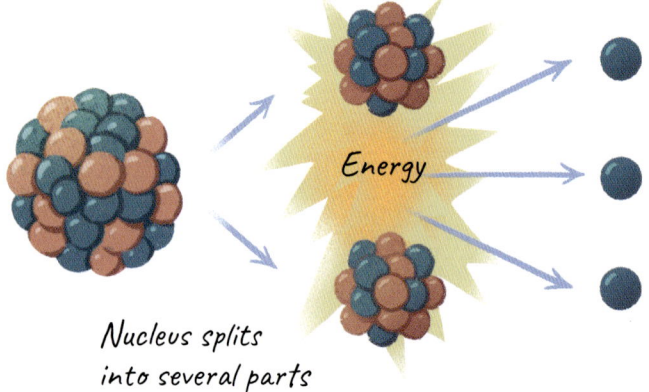

Nucleus splits into several parts

Fission and fusion occur naturally, but we can also make them happen deliberately.

In **nuclear fusion**, much smaller atoms, such as hydrogen atoms, combine their nuclei, making new atoms. The new atom needs less energy to hold itself together than the two separate atoms, so energy is released.

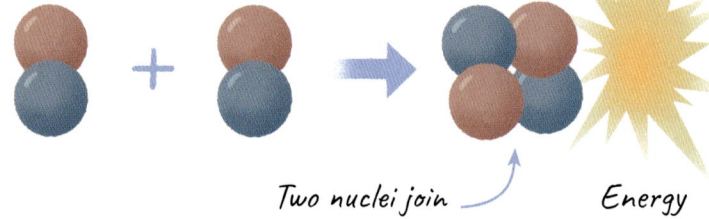

Two nuclei join — Energy

Splitting the atom

Scientists first figured out how to split an atomic nucleus deliberately in the 1930s. They fired proton at a lithium atom very fast, smashing it apart.

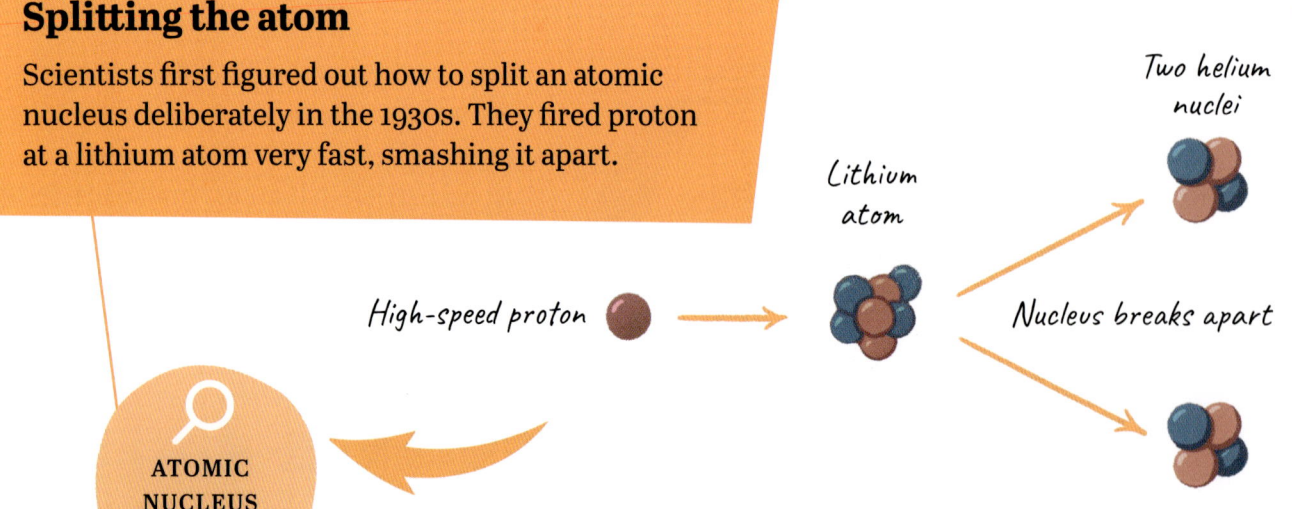

High-speed proton → Lithium atom → Two helium nuclei / Nucleus breaks apart

ATOMIC NUCLEUS PAGE 68

Nuclear weapons

Nuclear fission was used to invent the first atom bombs, or nuclear weapons.

In an atom bomb, one atom is split apart, releasing particles that split more atoms apart, and so on. This leads to a chain reaction and a big explosion.

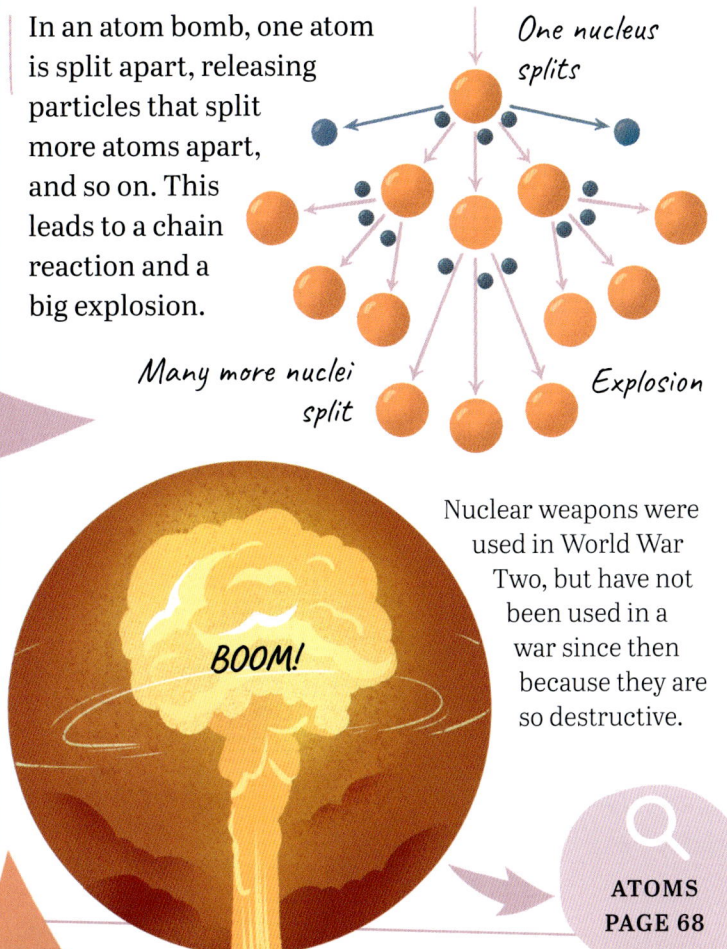

One nucleus splits

Many more nuclei split

Explosion

Nuclear weapons were used in World War Two, but have not been used in a war since then because they are so destructive.

BOOM!

🔍 ATOMS PAGE 68

Nuclear power

We can also use fission as a source of useful energy in nuclear power plants. Instead of an uncontrolled explosion, nuclear power works by splitting uranium atoms in a controlled way. The heat this releases drives turbines to generate electricity.

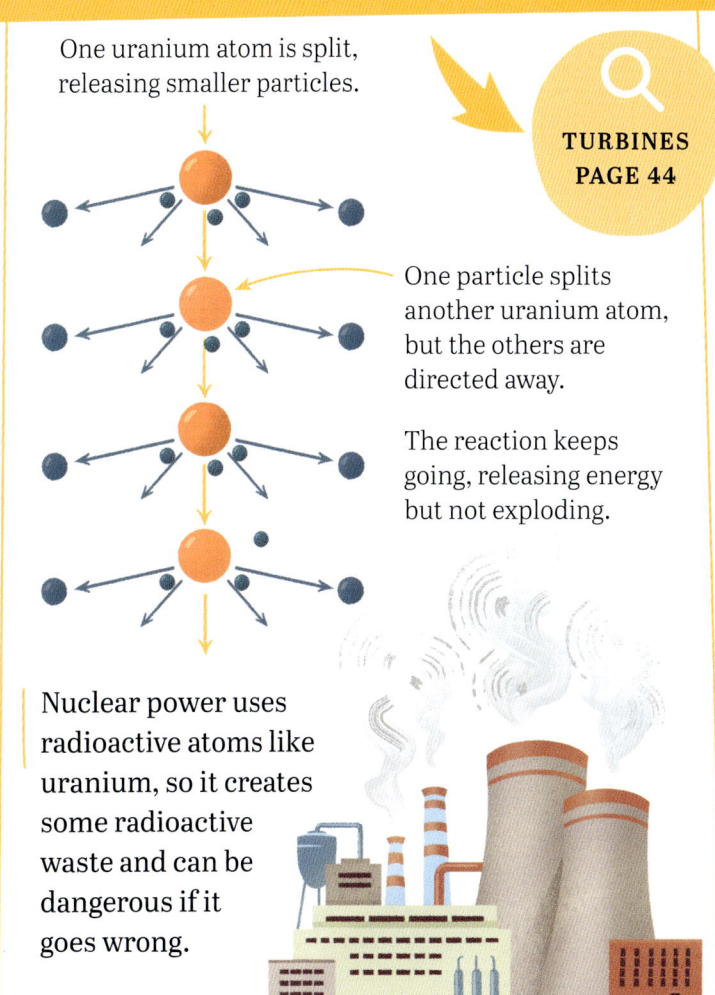

One uranium atom is split, releasing smaller particles.

One particle splits another uranium atom, but the others are directed away.

The reaction keeps going, releasing energy but not exploding.

Nuclear power uses radioactive atoms like uranium, so it creates some radioactive waste and can be dangerous if it goes wrong.

🔍 TURBINES PAGE 44

Nuclear fusion

Nuclear fusion is the opposite of nuclear fission. It happens naturally in the Sun and other stars, where atoms of hydrogen smash and fuse together, creating helium atoms. This releases vast amounts of heat and light, which is why the Sun is so hot and bright.

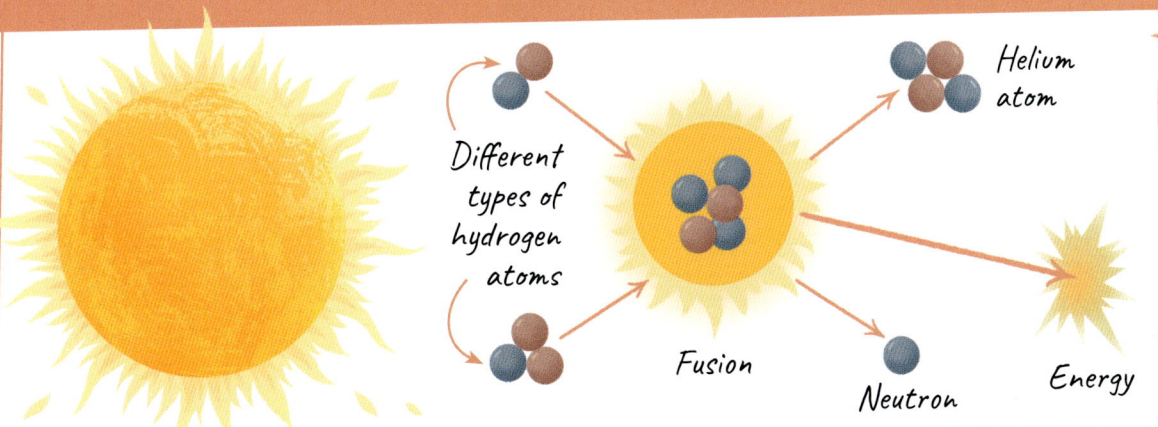

Different types of hydrogen atoms

Fusion

Helium atom

Neutron

Energy

Scientists are working on ways to use fusion for nuclear power instead of fission. If we can do this, it should be cleaner and safer than fission power.

🔍 HEAT ENERGY PAGE 22

Mysteries of matter

Atoms are made of smaller parts, or particles. But what are those particles made of ... and what actually IS matter?

Scientists are still trying to find out more about matter and how it works.

The parts of an atom are known as subatomic particles. Each type of atom has different numbers of subatomic particles. They are held together by pulling forces inside the atom.

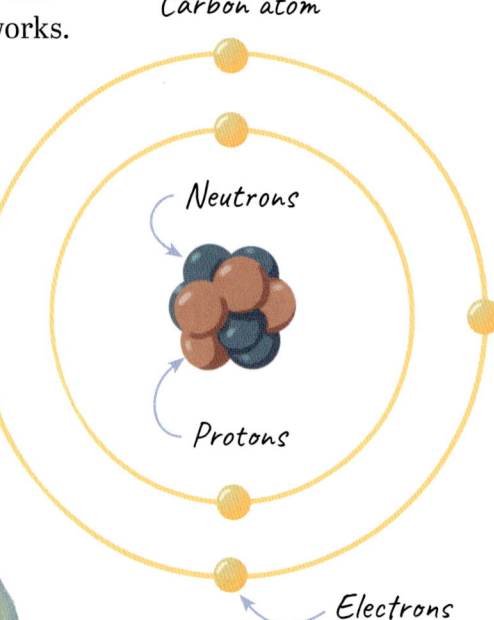
Carbon atom
Neutrons
Protons
Electrons

Fundamental particles are so small, it's very hard to find them, but scientists are learning more and more about them.

Neutron

Fundamental particles

Protons and neutrons are themselves made of even tinier particles called fundamental particles.

Looking inside

Physicists find out about fundamental particles using machines called particle accelerators. They smash subatomic particles together at an incredibly high speed. This breaks them up into smaller parts, which the scientists can detect and study.

The Large Hadron Collider (LHC) in Switzerland is one of the world's biggest particle accelerators. It's a ring-shaped tunnel 8 km (5 mi) across, built underground.

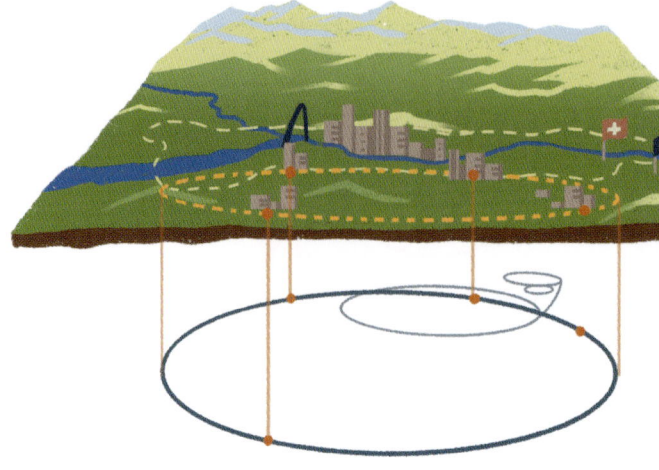

🔍 PHYSICISTS PAGE 8

Powerful magnets drive particles around the ring extremely fast.

76

The fundamental particles

Physicists have discovered at least 30 different fundamental particles, and have given them all names. They include the electron, and a whole range of others.

🔍 NEUTRONS PAGE 68

They've also discovered how fundamental particles work together to make bigger subatomic parts, such as neutrons.

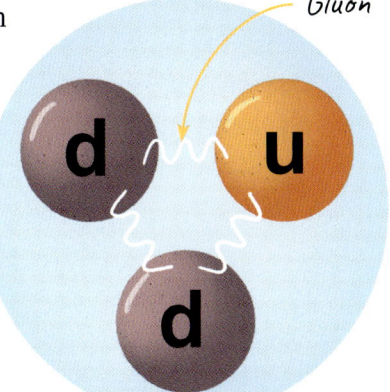

For example, a neutron is made of:
- One up quark
- Two down quarks
- Three gluons, which hold the quarks together

What are fundamental particles made of?

The smaller the particles are, the harder it is to study them. We can detect fundamental particles and track their movements, but we can't see what they are made of. There are theories about how they work, but no one knows for sure.

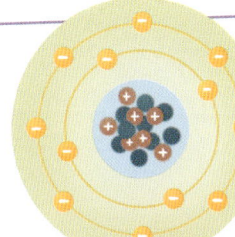

One theory, called string theory, says that each tiny fundamental particle is like a vibrating loop of energy.

🔍 ENERGY PAGE 16

Matter and energy

Scientists have also found that matter can turn into energy, and energy can turn into matter. The great physicist Albert Einstein studied this. He came up with a famous equation that shows how much energy matter can turn into.

$E = mc^2$ means that the energy in matter (E) equals the mass of the matter (m) multiplied by the speed of light (c) multiplied by itself (squared, or ^{2}).

It's hard to understand, but it basically means that a small amount of matter is equal to a LOT of energy!

Geophysics

Geophysics is the physics of the Earth and its processes.

Geophysicists study what the Earth is made of and the energy, movements, and forces at work inside it, on its surface, and in its atmosphere (the gases surrounding it).

The study of geophysics includes:
- The structure of the Earth
- The Earth's gravity and magnetic field
- The atmosphere
- Ocean zones, waves, and currents
- How landforms are made
- The age of the Earth, rocks, and fossils
- How earthquakes and volcanoes happen

Landforms are the different shapes and features of the landscape, such as bays, mountains, cliffs, and valleys.

Though the Earth usually feels strong and solid to us, both its surface and the parts inside it are constantly moving and changing.

Inside the Earth

We can't explore deep inside the Earth to study it, because it's too hot. Instead, geophysicists study things like how vibrations move through the layers of the Earth, to figure out what each layer is made of.

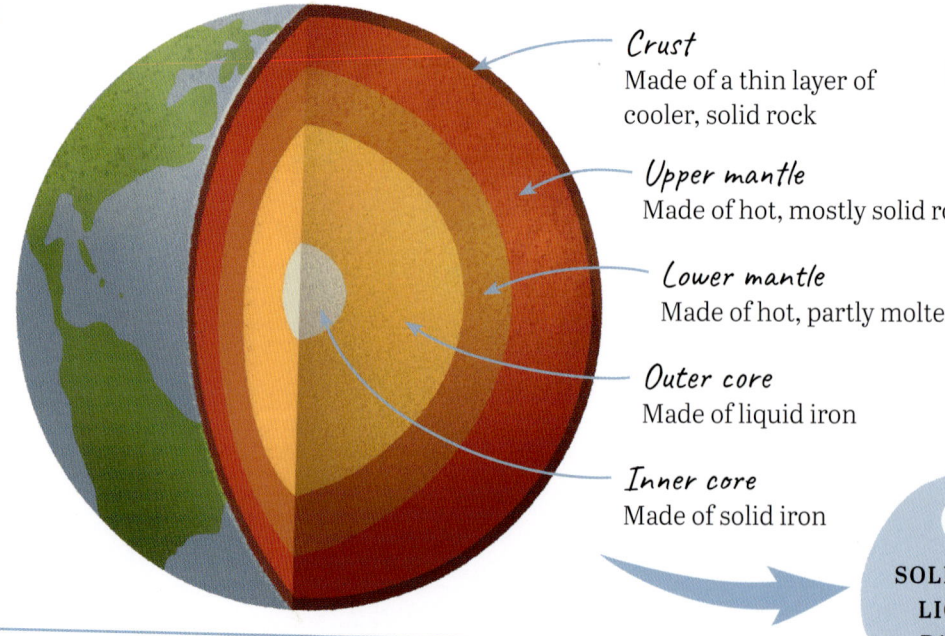

Crust — Made of a thin layer of cooler, solid rock

Upper mantle — Made of hot, mostly solid rock

Lower mantle — Made of hot, partly molten rock

Outer core — Made of liquid iron

Inner core — Made of solid iron

SOLIDS AND LIQUIDS PAGE 70

Magnetic Earth

The Earth is a giant magnet and has a magnetic field stretching out into space around it. Scientists think its magnetism is caused by the way the liquid iron in the outer core moves around.

North magnetic pole
The Earth
Magnetic field
South magnetic pole

The Earth's magnetic field protects us from high-energy particles and waves that reach the Earth from space or from the Sun. It makes them flow around the Earth instead of hitting it directly.

Sometimes, the particles make gases in the atmosphere glow around the North or South Pole. We can see this as glowing light in the sky, called an aurora.

🔍 MAGNETISM PAGE 60

Tectonic plates

The Earth's crust is not all in one piece. Instead, it's made up of giant sections, a bit like huge jigsaw puzzle pieces, called tectonic plates.

Tectonic plate
Plate boundary, or edge

In some places, new crust forms as molten rock from inside the Earth pushes upward and spreads out.

In others, plates push or scrape together, or one slides underneath another.

🔍 MELTING ROCK PAGE 71

Natural disasters

Over millions of years, the moving plates make the continents move around and change shape. Where the plates meet, their movements can cause earthquakes, volcanic eruptions, and tsunamis. Geophysicists can sometimes help to save lives by tracking and predicting natural disasters like these.

When two plates push together and then suddenly slip, it can cause an earthquake.

An undersea quake can make the sea move suddenly, creating a tsunami wave.

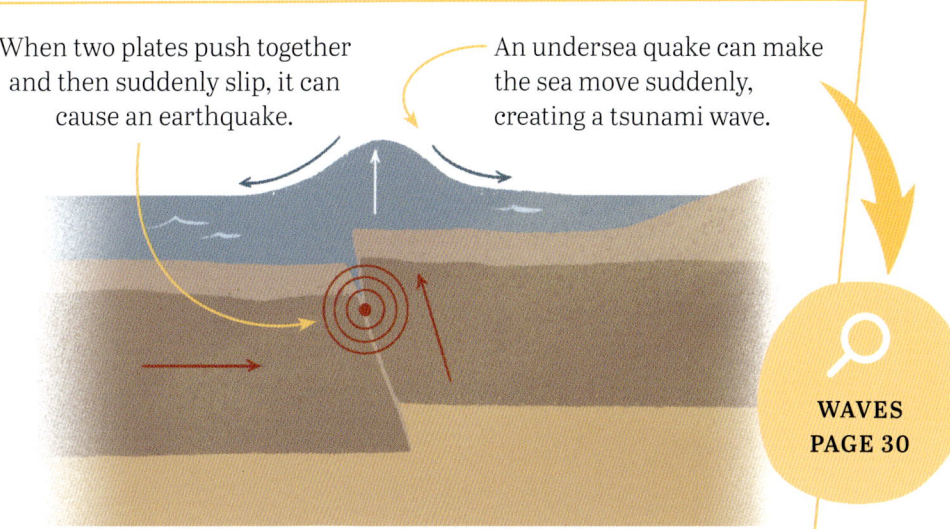

🔍 WAVES PAGE 30

The Universe and space

Our planet, the Earth, is part of a vast Universe stretching away from us in every direction.

Physicists who study the Universe and space are known as astrophysicists or cosmologists.

The **Universe** means everything that exists—all the stars, planets, moons, asteroids, comets, space dust, and everything in between, including the Earth.

The word *Universe* means "everything in one."

Space means everything beyond the Earth and its atmosphere.

The Universe is enormous, and we have only been able to explore a tiny area of it close to the Earth. But physicists can study it using telescopes, and find clues about how it works.

When rockets and astronauts go to space, it means they travel at least 100 km (62 mi) away from the Earth's surface, past an invisible boundary known as the Kármán Line.

Expanding Universe

By studying the stars, scientists have found that the Universe is expanding, or getting bigger. Space objects seem to be moving apart and spreading out.

Galaxies in outer space are moving away from each other.

GALAXIES PAGE 83

The Big Bang

Because the Universe appears to be expanding, many scientists think it could have started by suddenly growing out from a tiny single point. This theory is known as the Big Bang.

This diagram shows what scientists think happened:

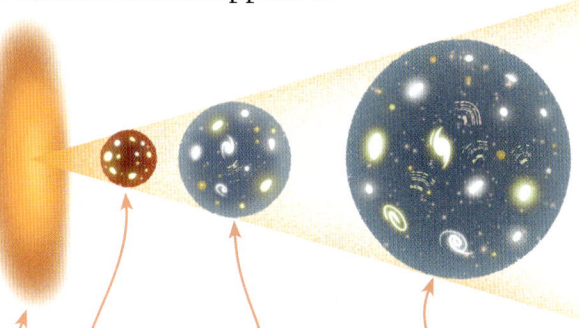

13.8 billion years ago — The Big Bang, when the Universe started.

3 minutes later — The first nuclei of atoms formed.

100 million years after the Big Bang — The first stars formed.

400 million years after the Big Bang — The first galaxies (giant clusters of billions of stars) formed.

9.2 billion years after the Big Bang (4.6 billion years ago) — The Sun and solar system formed, including the Earth.

🔍 **STARS PAGE 82**

Light-years

Light-years are used to measure distances in space.

One light-year is the distance that light travels in one year. Since light travels extremely fast, that's a huge distance: about 9.46 trillion km (5.8 trillion mi).

One year is one orbit of the Earth around the Sun.

A light-year is how far light travels in that time.

4.2 light-years

106,000 light-years

Our galaxy, the Milky Way, is about 106,000 light-years across.

The nearest star to our Sun, Proxima Centauri, is about 4.2 light-years away.

🔍 **ORBITS PAGE 84**

Back in time

When we look at faraway space objects, we are also looking back in time.

For example, the star Rigel is 864 light-years away from Earth.

Orion constellation

Rigel is visible in the night sky as part of the constellation Orion.

Rigel

We see Rigel because light from it reaches our eyes.

Since it's 864 light-years away, the light has taken 864 years to get here.

So we are seeing Rigel as it was 864 years ago!

🔍 **TIME PAGE 11**

Shining stars

Humans have always looked up at the stars and wondered what they were. Now we know much more about them, thanks to powerful telescopes.

Stars are balls of gas that are constantly exploding, releasing a lot of light and heat. There are thought to be about 200,000,000,000,000,000,000,000 stars in the Universe—that's 200 thousand million million million!

Sun

Our Sun is a star—the closest one to us.

The Sun is a medium-sized star, about 1.4 million km (865,000 mi) across—much bigger than the Earth.

← *Earth to scale*

Astrophysics means the physics of stars. Astrophysicists study how stars form and change over time, and the different types of stars.

How stars begin

Stars are born in big clouds of gas and dust in space called nebulas. Gas and dust whirl and clump together, pulled by their own gravity, and form a dense ball. Eventually, it gets so big, heavy, and hot that nuclear fusion begins, and the star starts to give out heat and light.

Nebula

New stars

🔍 **NUCLEAR FUSION PAGE 75**

Life of a star

Stars are not alive, but physicists describe them as having a "lifetime" made up of several stages. This diagram shows the lifetime of an average Sun-sized star, which can exist for around 10 billion years.

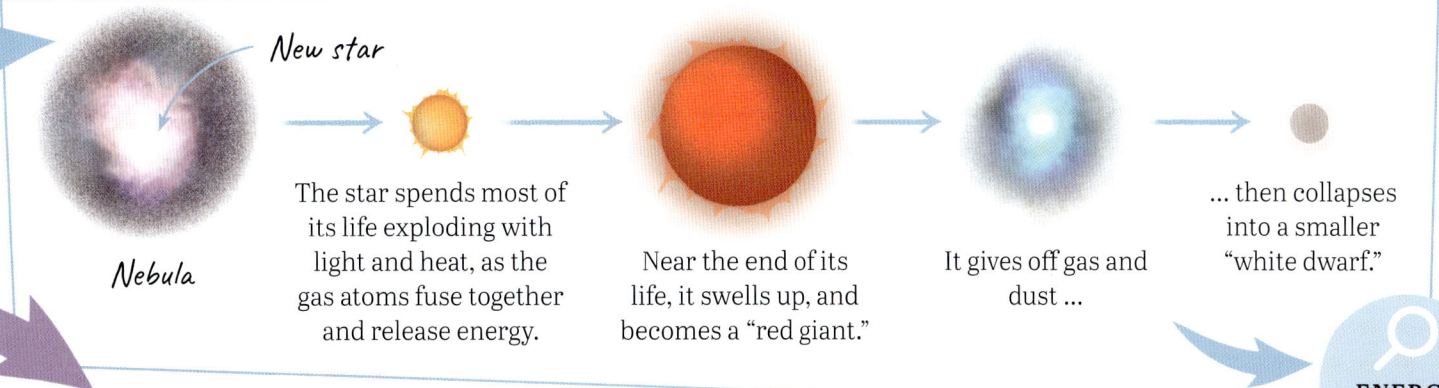

Nebula — New star

The star spends most of its life exploding with light and heat, as the gas atoms fuse together and release energy.

Near the end of its life, it swells up, and becomes a "red giant."

It gives off gas and dust ...

... then collapses into a smaller "white dwarf."

🔍 ENERGY PAGE 16

Galaxies

A galaxy is a ginormous cluster of stars, spinning around in space and held together by gravity. The Sun and the planets orbiting around it are part of our home galaxy, the Milky Way. We have also spotted many other galaxies, far away in outer space, and have given them names.

Whirlpool galaxy

Tadpole galaxy

Cartwheel galaxy

🔍 GRAVITY PAGE 52

Strange stars

Space scientists sometimes discover new and unusual types of stars.

🔍 ORBITS PAGE 84

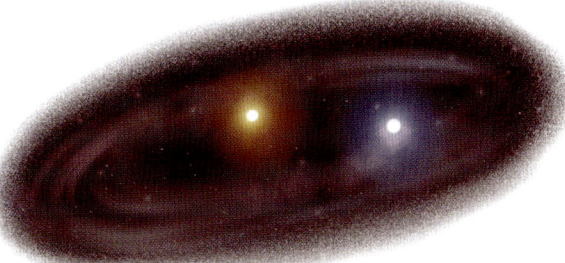

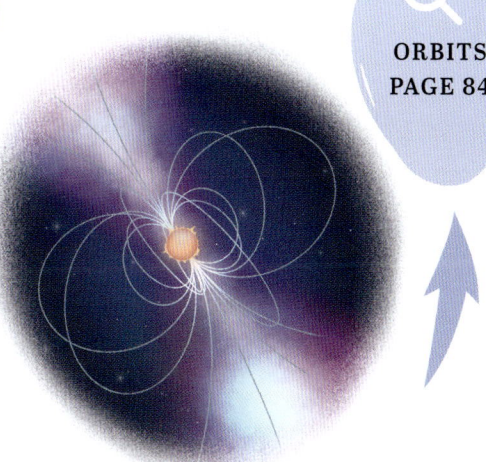

A supernova is a massive explosion that can happen when a very large star runs out of fuel.

A binary star looks like one star in the night sky, but it is actually two stars orbiting closely around each other.

Pulsars are tiny, very dense stars that spin incredibly fast—some rotate more than 700 times per second.

83

In orbit

The Universe is full of objects orbiting around other objects—but why?

Orbiting means moving around and around another object in a curved path. For example, the Earth orbits the Sun, and the Moon orbits the Earth.

The planets orbit the Sun.

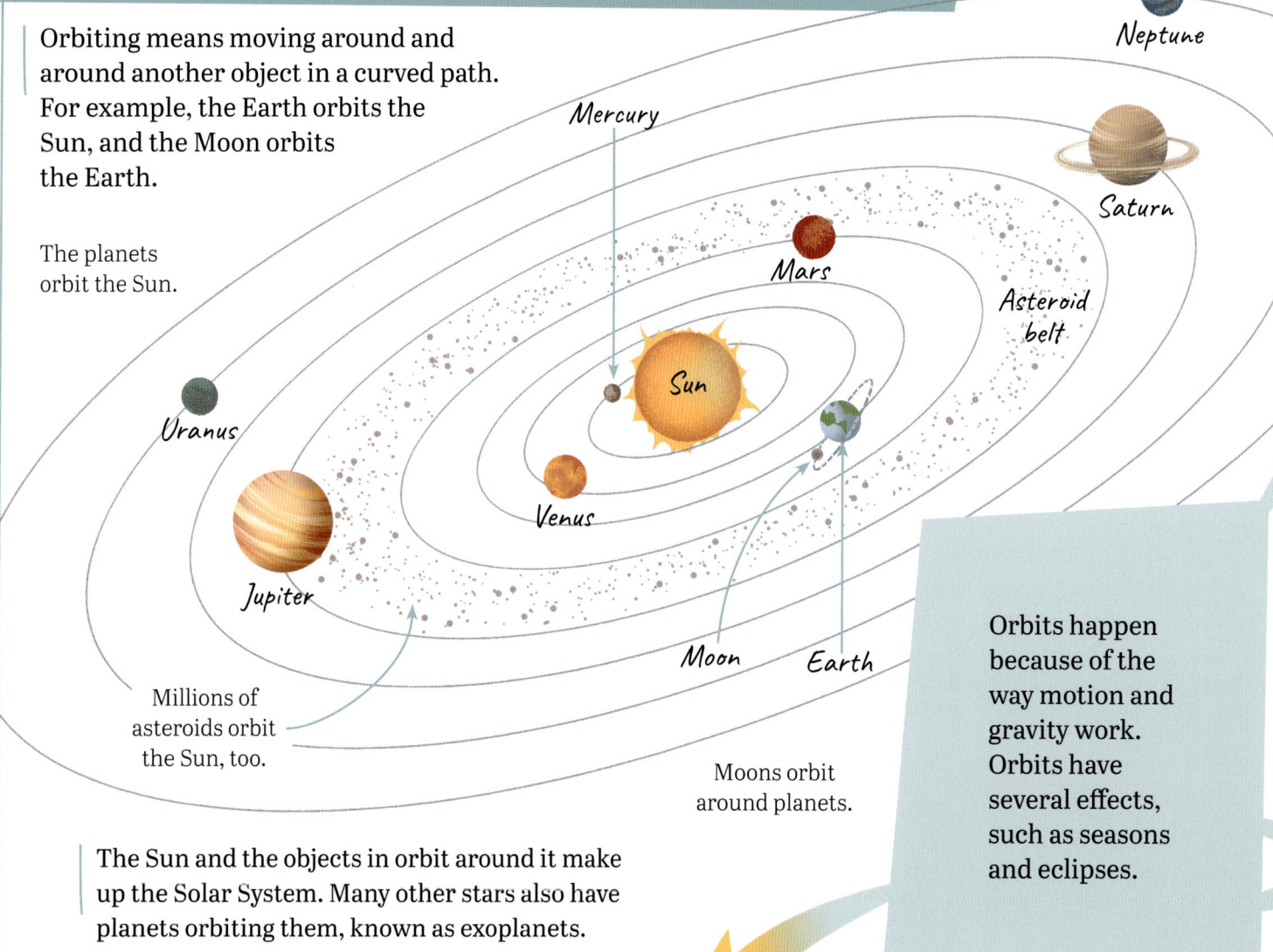

Millions of asteroids orbit the Sun, too.

Moons orbit around planets.

Orbits happen because of the way motion and gravity work. Orbits have several effects, such as seasons and eclipses.

The Sun and the objects in orbit around it make up the Solar System. Many other stars also have planets orbiting them, known as exoplanets.

How orbits work

This diagram of the Earth and the Moon shows how orbits work.

The Moon is zooming through space. As Newton's first law states, it will keep moving at the same speed and in the same direction unless another force acts on it.

The Earth's gravity pulls on the Moon, but it's not strong enough to pull the Moon down to the ground.

Instead, the Moon's forward motion and the pull of gravity are in balance. The Moon constantly curves around the Earth, but never touches it.

🔍 **NEWTON'S FIRST LAW PAGE 62**

84

The seasons

Orbits give us years and seasons. A year is one complete orbit of the Earth around the Sun.

The Earth is slightly tilted to one side. As it orbits, different parts of it lean toward or away from the Sun, making the hours of daylight and temperature change through the year—especially nearer to the poles.

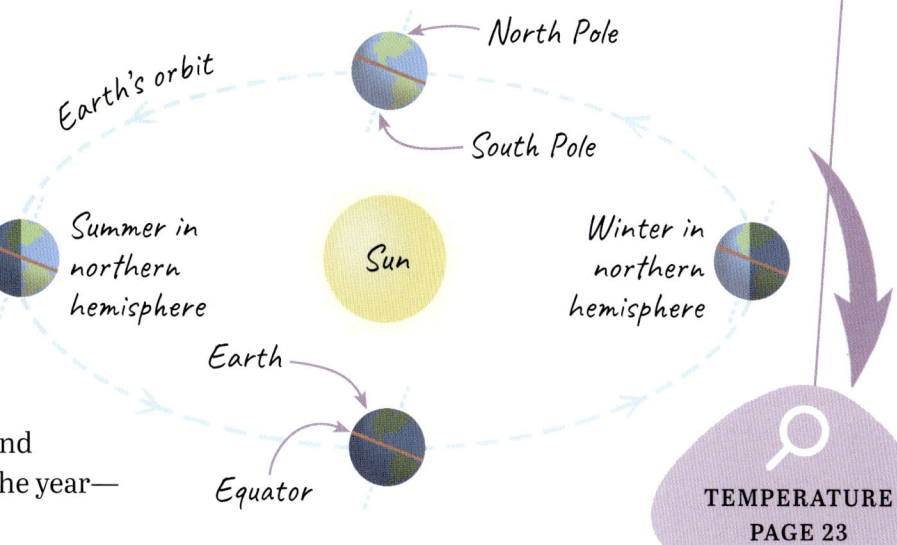

TEMPERATURE PAGE 23

Eclipses

Sometimes, the Moon orbits the Earth, just as the Earth, Moon, and Sun line up, causing an eclipse.

A solar eclipse, or eclipse of the Sun, happens when the Moon is between the Sun and the Earth.

The Moon blocks the Sun and casts a shadow on the Earth.

A lunar eclipse, or eclipse of the Moon, happens when the Earth is between the Sun and the Moon.

The Earth blocks the Sun and casts a shadow on the Moon.

SHADOWS PAGE 36

Artificial orbits

Planets and moons orbit naturally, but we can also send human-made spacecraft and satellites into orbit around the Earth, Moon, Sun, or other planets.

Smaller satellites do jobs like tracking weather patterns, sending phone signals, or watching space.

The International Space Station orbits the Earth.

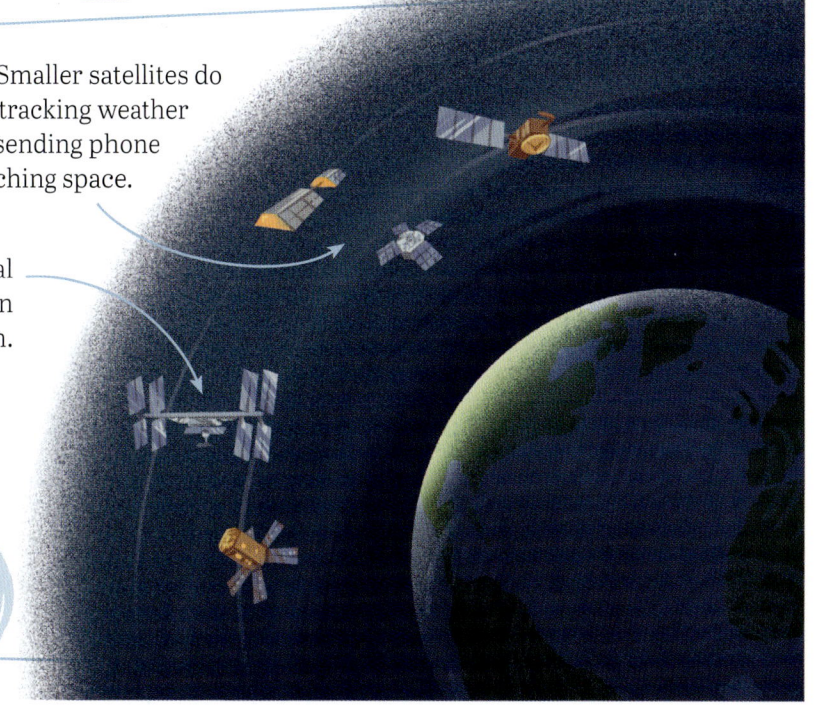

SPACE TELESCOPES PAGE 87

Studying space

To study space, the Moon, the stars, faraway galaxies, and how the Universe works, astrophysicists use a lot of hi-tech equipment.

People have been studying the stars and planets since ancient times. But for most of history, we could only observe what we could see with our own eyes.

By 3,000 years ago, the ancient Babylonians were studying the stars and predicting the movements of the planets. Around 1,000 years ago, Arabic and Persian scientists wrote books about the stars and gave many of them their names.

For the last 400 years, we've been able to look at space in more detail, thanks to the invention of telescopes. More recently, we've also sent spacecraft and astronauts into space to explore.

Persian scientist Abd al-Rahman al-Sūfī, who lived in the 900s, wrote about about dozens of constellations and their stars.

A closer look

Telescopes were first invented around 1600, for seeing long distances during battles. But people soon started using them to look at the night sky, including the great physicist Galileo.

Galileo Galilei

GALILEO PAGE 64

Types of telescopes

Early telescopes were all made with glass lenses, but today there are several types.

- Optical telescopes collect light from distant objects using lenses or mirrors.
- Radio telescopes detect radio waves coming from stars and other space objects.
- Other telescopes collect gamma rays, microwaves, or X-rays.
- Some space telescopes are on Earth, and others are satellites in space.

INFRARED LIGHT PAGE 38–39

The James Webb Space Telescope is a powerful space telescope that was launched into orbit in 2021. It collects infrared light.

Space travel

In the 1940s, humans began building rockets powerful enough to leave the Earth and fly into orbit. Russian cosmonaut Yuri Gagarin became the first person in space in 1961, and in 1969 people landed on the Moon for the first time.

On space stations orbiting the Earth, astronauts experiment with weightlessness and test the effects of space on the human body, and on other living things.

WEIGHT PAGE 53

Astronauts explored the Moon and collected rocks and dust to study back on Earth.

Uncrewed missions

To explore farther away than humans can travel, we use spacecraft with no one on board.

Space probes are small spacecraft sent to explore space and send information back to Earth. They've visited planets and moons throughout the Solar System, as well as the Sun itself.

Landers are probes that can land on planets, moons, or other space objects, and rovers are robots that can move around and explore on the surface.

THE SUN PAGE 82

The Parker Solar Probe is orbiting the Sun, and has flown closer to it than any other spacecraft.

Sojourner was the first rover to land on and explore the planet Mars.

87

Mysteries of the Universe

There are many things about space and the Universe that we don't yet understand ...

To study things like orbits, gravity, and stars, physicists use the same laws of physics that they apply to things on Earth.

For example, stars behave the way they do because of the nuclear fusion inside them. We can measure their light and heat because these things work the same way all over the Universe.

Stars have different temperatures, which can make them appear different colors.

But some things in space don't work as we expect them to, or seem to break the laws of physics ...

Black holes

Black holes can form when a giant star runs out of fuel. It collapses inward and becomes more and more dense. The denser it gets, the more it shrinks. Eventually, it becomes a single point in space that contains a huge amount of mass, but has no size at all!

Black holes have powerful gravity and pull in anything that gets close to them—even light.

Event horizon

You can't see a black hole, but it has a bigger zone around it called the event horizon. Inside this, nothing can escape from the black hole, so it looks dark. We can often see stars and dust swirling around the event horizon, getting pulled in.

MASS PAGE 53

GIANT STARS PAGE 83

Dark matter

Scientists have found that stars, galaxies, and other space objects don't always move around in the way they expect. Instead, they act as if they are being affected by gravity from invisible matter. They have named this "dark matter." It could be made of some kind of black hole or something else that we don't understand yet.

🔍 GRAVITY PAGE 52

Do aliens exist?

As far as we know, life only exists on Earth. But there are so many other stars, planets, and galaxies, that scientists think life should have developed in lots of places. So why haven't we seen or heard from any aliens?

There are several theories:
- They do exist, but they are too far away to contact us.
- They do exist, but they are basic life forms that don't have the technology to contact us.
- There's no life anywhere apart from on Earth—but no one knows why.

Movies often show alien spacecraft looking like this, but no one knows how they might really look.

🔍 GALAXIES PAGE 83

Ideas about the Universe

The Big Bang is the main theory about how the Universe began, but there are other ideas, too, such as:

- The Universe has no start or end, but just keeps growing and shrinking.
- The Universe only appears as it does to us because we exist in 3D. It has many other dimensions, but we can't experience them.
- The Universe is just one of countless universes in an endless "multiverse."

🔍 THE BIG BANG PAGE 81

Physics super mind map

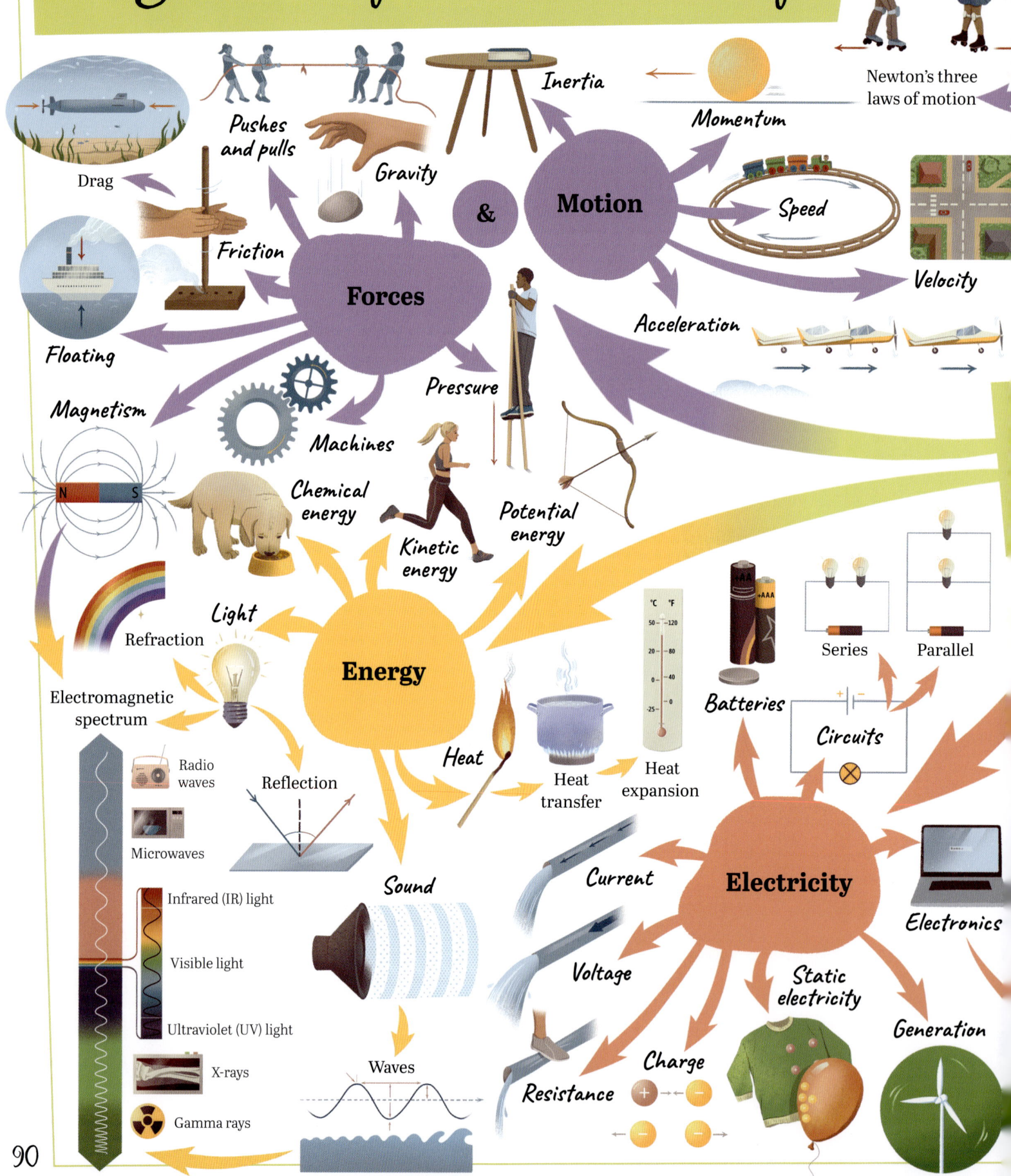

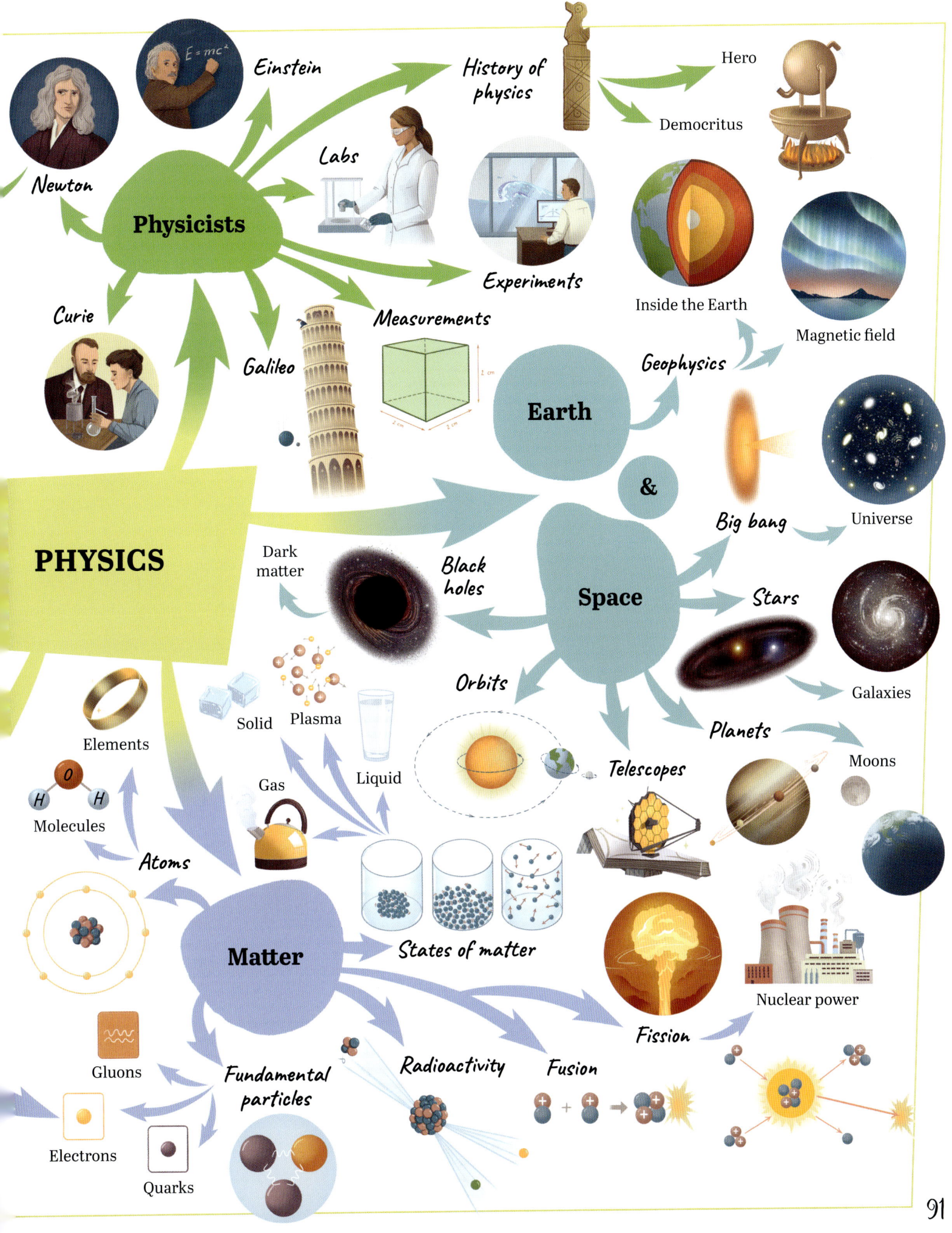

Glossary

Absolute zero The lowest possible temperature that could exist, if atoms and molecules did not move at all.

Acceleration An increase in speed.

Air pressure Another name for atmospheric pressure.

Air resistance A type of friction that slows an object's movement through the air.

Atmospheric pressure Pressure on the Earth's surface from the atmosphere, the layer of gases surrounding the Earth.

Atoms The tiny units that make up all matter.

Battery A store of chemical energy that can provide a supply of electricity.

Big Bang A theory that the Universe began with a sudden expansion of all matter and energy from a tiny single point, around 14 billion years ago.

Binary code A code made up of 0s and 1s, used in electronic devices to store information and do calculations.

Black hole An invisible point in space with powerful gravity, which can form after an old star collapses.

Carbon dioxide (CO_2) A gas found in the air and produced as waste when fuels burn.

Celsius A temperature scale often used in science, with the freezing point of water at 0 °C and the boiling point of water at 100 °C.

Charge A property of matter that can be positive or negative, and can make it experience a force when in a magnetic or electric field.

Chemical energy Energy stored in chemical substances, such as food or fuel.

Chemical reaction A process in which atoms rearrange themselves to form new materials when substances are combined.

Chip See *Silicon chip*.

Circuit A loop of metal wire or another conductor, which electricity can flow around.

Climate change A long-term change in weather and climate patterns.

Compound A material made from atoms of different elements joined together to form molecules.

Condense To change from a gas into a liquid.

Conduction Heat moving through an object, or from one object to another object that is touching it.

Conductor A material that is good at letting heat or electricity flow through it.

Conservation of energy A law of physics which states that energy cannot be created or destroyed, only converted from one form to another.

Constellation A group of stars that form a pattern in the sky when viewed from Earth.

Convection The movement of heat through a liquid or gas caused by warmer atoms or molecules moving around.

Core The middle part of the Earth, deep below the surface.

Crust The rocky surface layer of the Earth.

Current A flow of electricity around a circuit or through a material.

Density The weight of an object compared to its volume (the amount of space it fills).

DNA (short for **Deoxyribonucleic Acid**) A chemical found in cells, containing coded instructions that make living things work.

Drag A type of friction that slows an object's movement through a gas or liquid.

Echo A sound caused by the reflection of sound waves from a surface.

Elastic Able to stretch or change shape and then return to its original size or shape.

Electricity A form of energy made from a flow or buildup of electric charge or charged particles, such as electrons.

Electromagnetic (EM) waves Energy waves made of ripples in electric and magnetic fields, which can travel through a vacuum or empty space.

Electromagnetic spectrum (EMS) The spectrum or range of different wavelengths of electromagnetic waves, including X-rays, radio waves, and visible light waves.

Electronics The area of science and technology that uses the flow of electricity in a circuit to store information and do calculations in, for example, computers and smartphones.

Electrons Tiny particles that zoom around the nuclei of atoms.

Elements Pure materials made of only one type of atom, such as gold or carbon.

Energy The power to do work and make things happen.

Evaporate To change from a liquid to a gas.

Fahrenheit A temperature scale that has the freezing point of water at 32 °C and the boiling point of water at 212 °C.

Field An area that is affected by a force such as electrical charge, magnetism, or gravity.

Fission See *Nuclear fission*.

Force A push or pull that can make an object change speed, shape, or direction.

Freeze To change from a liquid into a solid by getting colder.

Friction A force that slows moving objects down or gives objects grip against each other.

Fuel A store of chemical energy used to power machines or other devices.

Fundamental particles The smallest possible particles of matter, which scientists think make up the subatomic particles found in atoms

Fusion See *Nuclear fusion*.

Galaxy A huge, rotating cluster of millions or trillions of stars.

Gamma rays High-energy electromagnetic waves given off by some radioactive atoms.

Gas A state of matter in which atoms or molecules move around freely and spread out to fill the available space.

Generator A machine that converts another form of energy, such as chemical or kinetic energy, into an electricity supply.

Global warming A gradual increase in Earth's average temperature over the last two centuries, caused by human activities.

Gravity A force found in all objects with mass, which pulls other objects toward them.

Heat expansion Expanding or getting bigger with increased temperature.

Heat transfer The flow of heat from place or object to another.

Inertia The way objects keep doing the same thing (such as moving or staying still) unless a force acts on them to change that.

Infrared (IR) light A type of light energy given off by hot objects, with a wavelength too long for humans to see.

Insulator A material that does not let heat or electricity flow through it easily.

Ion An atom that has extra or missing electrons, giving it a negative or positive charge.

Kelvin A temperature scale that starts at absolute zero.

Kinetic energy Movement or motion energy that an object has when it is moving.

Lander Part of a spacecraft that lands on a planet, moon, or other space object.

Lift Upward pressure on an object caused by a liquid or gas moving around it.

Light-year The distance light travels in one Earth year, used as a measurement of distances in space.

Liquid A state of matter in which atoms or molecules move around loosely, allowing it to flow, drip, or fill a container.

Magnet A material or object that produces a magnetic field, containing a force that can attract some types of metals and attract or repel (push away) other magnets.

Mantle A thick layer of rock, some molten and some solid, inside the Earth between the crust and the core.

Mass The amount of matter there is in an object or substance.

Materials Different types of matter.

Matter The stuff that everything is made up of.

Microwaves A type of electromagnetic wave with a longer wavelength, used in microwave ovens to heat food.

Mixture A material made of different elements or compounds mixed together.

Molecules Units of matter made from atoms bonded (joined) together.

Momentum A measurement of the combined mass and velocity of an object.

Motion Another name for movement.

Motor A device that turns fuel or electricity into motion to make a moving part of a machine work.

Neutron A type of subatomic particle found in the nucleus of an atom.

Nuclear fission Breaking apart the nucleus of an atom to release energy.

Nuclear fusion Fusing or joining the nuclei of atoms to release energy.

Nuclear power A supply of electricity or other energy made using nuclear fusion or fission.

Nucleus The central part of an atom.

Orbit A regular, repeating path of one object in space around another.

Particle A tiny part, such as the parts that make up atoms.

Photosynthesis The process of using energy from the Sun to convert water and carbon dioxide gas into glucose, inside a plant's leaves.

Physicist A scientist who studies physics.

Plasma A state of matter in which electrons are free from atoms. It's found in lightning, flames, and the Sun and other stars.

Poles The two ends of a magnet, known as north or south, where its magnetic force is strongest.

Potential energy Energy that has the potential to become another form of energy, such as the energy stored in a stretched spring or a ball at the top of a slope.

Power The rate at which energy is used or converted into another form of energy.

Properties Features or characteristics of something, such as a material or a particle.

Proton A type of subatomic particle found in the nucleus of an atom.

Quarks One of the types of tiny fundamental particles that are thought to form matter.

Radiation Another name for electromagnetic energy waves, such as light or radio waves.

Radioactivity The release of energy, in the form of waves or moving particles, from the nuclei of some types of atoms.

Radio waves A type of electromagnetic wave with a longer wavelength, often given out by objects in space, and used to carry information over long distances.

Reflection A change in direction of an energy wave, such as light or sound, when it hits a surface.

Refraction A change in direction of an energy wave, such as light or sound, when it passes from one material into another.

Renewable energy Energy from sources that do not run out, such as wind, waves, and sunshine.

Resistance A part or property of an electric circuit that slows down the flow of current.

Rover In space exploration, a robotic vehicle that can explore the surface of a faraway planet, moon, or other space object.

Silicon chip A tiny electronic circuit printed onto a small piece of the element silicon.

Solar panel A panel of material that turns energy from sunlight into a flow of electricity.

Solid A state of matter in which atoms or molecules are packed closely together and stay in a fixed shape.

Space probe A robotic spacecraft that explores space without any crew on board.

Speed The distance an object travels in a given amount of time.

Spring An object that can stretch or bend, storing energy, then return to its original shape.

States of matter The states that materials can exist in: solid, liquid, gas, and plasma.

Static electricity Electric charge that collects or builds up in or on an object instead of flowing as a current.

String theory A theory that matter is made of very small, stringlike vibrating loops.

Subatomic particles The tiny particles that make up atoms.

Tectonic plates The huge sections of rock that make up the Earth's crust.

Telescope A device that collects light, radio waves, or other types of radiation from space and uses them to create a magnified image.

Thermal To do with heat.

Thrust A pushing force that makes an object move.

Turbine A device that uses a movement such as a flow of steam to make a wheel spin, which is then used to generate electricity.

Ultraviolet (UV) light A type of light energy with a wavelength too short for humans to see.

Upthrust A force that pushes up on an object that is in a liquid or gas.

Vacuum A totally empty space, with no air or anything else in it.

Velocity Speed in a particular direction.

Visible light The range, or spectrum, of electromagnetic waves that can be seen by the human eye.

Voltage The difference in charge that makes electrons flow through a circuit.

Waves Regular, repeating patterns of vibration or disturbance that can travel through a material or through empty space.

Wavelength The length from a point on one wave to the same point on the next wave.

Weight The force caused by the pull of gravity on the mass of an object.

Work The amount of energy used or converted into another form of energy.

X-rays A type of high-energy electromagnetic wave with a shorter wavelenth.

Index

absolute zero 23
acceleration 64, 65
acoustic levitator 8
acoustics 33
air resistance 55, 62
al-Sūfī, Abd al-Rahman 86
aliens 89
alpha particles 72, 73
applied physics 9
area, measuring 10
astronauts 80, 86, 87
astronomers 13
astrophysicists 80, 82, 86–7
astrophysics 8, 82
atmosphere 78
atmospheric pressure 59
atom bombs 75
atoms 12, 13, 23, 40
　in electromagnetic waves 39
　in heat energy 22
　in heat expansion 26, 27
　in heat transfer 24–5
　in magnets 60
　making molecules 69
　in matter 68, 70
　radioactive 72–3, 74, 75
　splitting the atom 74
　types of 68, 76
aurora 79
axles 67

Babylonians 86
balloons 58
balls 21, 50, 56, 57, 62
batteries 28, 42, 45
bending 57
beta particles 72, 73
the Big Bang 81, 89
bikes 6, 7, 50, 55
binary code 49
binary stars 83
birds 64
black holes 9, 88
boats 55, 59
Bohr, Neils 13
bouncy castles 57
brass 69
bridges 14, 26
bulbs 42, 43
buzzers 42, 43

cables 46, 47
calories 29
carbon 69
cars 65
cats' eyes 37
cells, damaging 73
Celsius 23
charge 40, 60, 61
chemical energy 28
chips 48
circuits, electric 42–3, 48, 49
cobalt 60
colors, seeing 37
compression 57
computers 48, 49
conduction 24
conductors 25, 41, 47
contact forces 50
convection 24, 25
copper 6, 41, 69
cosmology 8, 80
crystals 37
Curie, Marie and Pierre 73
current 41, 43

dark matter 89
decibel scale 33
deep-sea submersibles 59
deformation 57
Democritus 12
digital information 49
distance, measuring 10
diving boards 21, 56
DNA, damage to 73
drag 51, 55
dynamics 63

$E = mc^2$ 77
Earth: atmospheric pressure 59
　eclipses 84, 85
　geophysics 78–9
　and gravity 52, 53, 62, 63, 78, 84
　inside the Earth 78
　magnetic Earth 79
　in orbit 84, 85
　tectonic plates 79
earthquakes 79
echos 33
eclipses 84, 85

Egyptians, ancient 12
Eiffel Tower 26
Einstein, Albert 77
elastic 21, 56–7
electricity 6, 7, 14, 40–1, 75
　electric circuits 42–3, 48, 49
　electric forces 60–1
　electric grid 46
　generating 44–5
　supplies 46–7
electromagnetic energy 31, 35, 72
　electromagnetic spectrum (EMS) 38–9, 72
electromagnetism 60, 61
electronics 48–9
electrons 40, 41, 68, 71, 72, 73, 77
elements 69
energy 6, 7, 16–17
　chemical energy 28
　electromagnetic 31, 35, 38–9, 72
　energy waves 30–1
　fission and fusion 74–5
　forms of 16
　heat energy 12, 14, 16, 22–3
　kinetic energy 16, 18–19, 20, 21, 22, 30, 44
　law of conservation of energy 17
　light energy 16, 30, 42
　and matter 17, 77
　potential energy 20–1
　renewable 45
　sound energy 16, 17, 30
　springs and elastic 21, 56–7
　see also electricity
equipment, measuring 11
event horizon 88
exoplanets 84
experiments 8
explosions 29, 75, 83

Fahrenheit 23
falling objects 53
fiber optic cables 37
fields 60
fireworks 29
fission 74–5
floating 59
Fomalhaut 13
food 16, 18, 28, 29
forces 7, 13, 50–1
　dynamics 63
　friction 7, 14, 50, 54–5, 62
　gravity 13–15, 45, 51, 52–3, 59, 62, 84

magnetic and electric 60–1
pressure 58–9
simple machines 66–7
friction 7, 14, 50, 54–5, 62
fuel 28, 44
fulcrum 67
fundamental particles 76, 77
fusion, nuclear 74–5, 82, 88

Gagarin, Yuri 87
galaxies 80, 81, 83
Galileo 13, 64, 86
gamma rays 39, 72, 73, 87
gases 7, 19, 25, 26, 57, 70
gears 67
generators 44, 46
geophysics 8, 78–9
global warming 44
glucose 69
gold 69
graphene 10
gravity 13, 14, 15, 45, 51, 52–3, 59, 62, 78, 84
black holes 88
dark matter 89
gravitational fields 53
force fields 53
mass and weight 53
Greeks, ancient 12

hairdryers 14
heat energy 12, 14, 16, 22–3
and friction 55
heat transfer 24–5
heat expansion 26–7
sensing 23
helium 68, 69, 75
Hero of Alexandria 12
hot-air balloons 27
hydroelectric power 21, 45
hydrogen 69, 74, 75

ice 71
inertia 62
infrared (IR) 25, 38
insulators 25, 41, 46, 47
inventions 15, 37
ionizing radiation 72, 73
iron 6, 41, 60, 69, 79
Islamic golden age 13

James Webb Space Telescope 87
joules (J) 17

Kármán Line 80
Kelvins (K) 23
kinetic energy 16, 18–19, 20, 21, 22, 30, 44

lab balance 11
landers 87
landforms 78
Large Hadron Collider (LHC) 76
law of conservation of energy 17
laws of motion 62–3, 84
Leaning Tower of Pisa 13
length, measuring 10
lenses 37
levers 67
lift 51
light 13, 34–5, 37
light energy 16, 30, 42
light waves 31, 35, 38, 72
reflection and refraction 36–7
spectrum 35, 37
speed of light 34
visible and invisible waves 38–9
light-years 34, 81
liquids 7, 19, 25, 26, 59, 70
lithium 74
load 43
lunar eclipses 85

machines, simple 66–7
magnetic forces 51, 60–1, 78, 79
mass 10, 19, 52, 53
matter 12, 15
dark matter 89
and energy 17, 77
and forces 7
heat energy 22
kinetic energy 19
mysteries of matter 76–7
states of matter 7, 70–1
temperature 23
waves in 30
what it is and how it works 6–7, 68–9
measurements 10–11, 12, 15
space and distance 10
time 11
units of 10, 11
metals 6, 25, 41, 56, 60
Michell, John 9
microwave ovens 15, 38, 87
Milky Way 83
mirrors 36, 37

mixtures, making 69
molecules 69, 70
momentum 65
the Moon 12, 13
eclipses 85
and gravity 52, 53, 63
humans on the 87
orbits 84
motion 84
laws of motion 62–3, 84
measuring 64–5
motors 42, 43
movement 16, 19
see also kinetic energy

natural disasters 79
neutrons 40, 68, 73, 76, 77
Newton, Isaac 13, 62–3, 84
Newtonian mechanics 8
North Pole 79, 85
nuclear fission and fusion 74–5, 82, 88
nuclear power 17, 75
nucleus 40, 68

oceanography, physical 8
optical telescopes 87
orbits 84–5
oxygen 68, 69

paper thickness 73
parachutes 63
Parker Solar Probe 87
particles: moving 19
particle accelerators 76
particle physics 8
photosynthesis 29
physicists 8–9, 14
planes 51, 55, 65
planets 12, 13, 16, 80, 84, 86–7
plants, chemical energy 29
plasma 71
plastic 25, 56
plug sockets 42, 47
poles 61
positrons 72
potential energy 20–1
power 17, 42, 75
power plants 44, 45, 46
pressure 58–9, 66
prisms 13
protons 40, 68, 73, 74, 76
pulsars 83

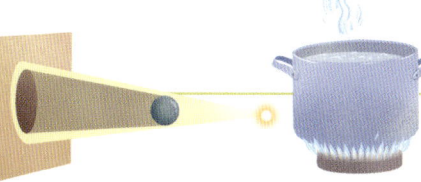

pylons 46

quantum mechanics 8

radiation 24, 25, 72–3
radio telescopes 87
radio waves 31, 35, 38, 72, 87
radioactivity 72–3
rainbows 35, 38
ramps 12, 66
red giants 83
reflection 35, 36–7
refraction 36–7
renewable energy 45
resistance 43, 55
Rigel 81
robots 49
rockets 28, 80
rocks 71
roller-coasters 15, 63
rovers 87
rubber 56
Rutherford, Ernest 13

safety 15
salt 69
satellites 85, 87
saucepans 6
scissors 66
seas 27, 52, 69
seasons 84, 85
seawater 69
semiconductors 48
shadows 36
silicon 48
silver 6, 41, 69
skyscrapers 26
smartphones 48, 49
smoke alarms 73
Sojourner 87
solar eclipses 85
solar panels 15, 45
Solar System 84, 87
solids 7, 19, 70
sound 32–3
  decibel scale 33
  sound energy 16, 17, 30
  sound waves 30, 32
South Pole 79, 85
space 9, 61, 80–1, 86–7
space stations 85, 87

spacecraft 85, 86, 87
speed: and laws of motion 63
  measuring 11, 64
Spencer, Percy 15
springs 21, 56–7
squeezing 57
stars 13, 16, 31, 34, 35, 71, 80, 81, 82–3, 88
  life of a star 82, 83
  nuclear fusion 75
  studying 86–7
states of matter 7, 70–1
static electricity 41
steam engines 12
streamlined shapes 55
stretching 57
subatomic particles 76, 77
substations, electricity 47
sugar molecules 28
the Sun 12, 25, 34, 75, 79, 81, 82, 83, 84, 85, 87
supernova 83
switches 43, 48, 49

tectonic plates 79
telescopes 13, 80, 82, 86, 87
temperature 23, 71
tension 57
theoretical physics 8, 9
thermal energy see heat energy
thermodynamics 8
thermometers 27
thrust 51
tides 45, 52
time 11, 15
towers 26
tsunamis 79
turbines 21, 44, 75
twisting 57

ultrasound 8
ultraviolet (UV) 39
the universe 80–1, 82, 84–5, 88–9
university physics 9
upthrust 59
uranium 73, 74, 75

vacuums 32, 34
velocity 65
vibrations 78
volcanoes 71, 79
voltage 43
volume, measuring 10

water 47, 69
  floating on 59

states of matter 7, 71
temperature 7, 23
water cycle 20, 21
water pressure 59
water resistance 55
waves 30, 31
watts (W) 17
waves 30–1, 35, 38–9
wedges 66
weight, and gravity 53
wheels 12, 67
white dwarves 83
wind power 44, 45
wires 41
wood 28
work 17
World War Two 75

X-rays 31, 35, 39, 72, 87

years 85